Intention and Intentionality

Essays in Honour of
G. E. M. Anscombe

Photo: Mary Gustafson

G. E. M. ANSCOMBE

Intention and Intentionality

Essays in Honour of G. E. M. Anscombe

EDITED BY
CORA DIAMOND
Associate Professor of Philosophy, University of Virginia

AND

JENNY TEICHMAN
Fellow in Philosophy, New Hall, Cambridge

CORNELL UNIVERSITY PRESS
Ithaca, New York

First published 1979 by Cornell University Press

Library of Congress Cataloging in Publication Data
Main entry under title:

Intention and intentionality.

1. Intention (Logic)—Addresses, essays, lectures.
2. Intentionalism—Addresses, essays, lectures.
3. Act (Philosophy)—Addresses, essays, lectures.
4. Necessity (Philosophy)—Addresses, essays, lectures.
5. Anscombe, Gertrude Elizabeth Margaret—Addresses, essays, lectures. I. Anscombe, Gertrude Elizabeth Margaret. II. Teichman, Jenny. III. Diamond, Cora.
BC199.I5I57 160 79–2478

ISBN 0–8014–1275–7

Photosetting by Thomson Press (India) Ltd., New Delhi
Printed in Great Britain by Redwood Burn Ltd., Trowbridge & Esher

CONTENTS

IV
TIME, TRUTH AND NECESSITY

LIST OF CONTRIBUTORS

J. E. J. ALTHAM

Lecturer in Philosophy in the University of Cambridge, Fellow of Gonville and Caius College, Cambridge. Author of *The Logic of Plurality.*

LUCY BROWN

Senior Lecturer in British History at the London School of Economics. Author of *The Board of Trade and the Free Trade Movement.*

BRIAN F. CHELLAS

Associate Professor of Philosophy at the University of Calgary. Author of *The Logical Form of Imperatives* and *Modal Logic: an Introduction.*

RODERICK M. CHISHOLM

Romeo Elton Professor of Natural Theology and Andrew Mellon Professor of Humanities at Brown University. His publications include *Person and Object: a Metaphysical Study, Realism and the Background of Phenomenology* and *Perceiving: a Philosophical Study.*

CORA DIAMOND

Associate Professor of Philosophy in the University of Virginia. Editor of *Wittgenstein's Lectures on the Foundations of Mathematics, Cambridge 1939.*

PETER GEACH

Professor of Logic in the University of Leeds. His publications include *Reference and Generality, Mental Acts, Logic Matters* and *God and the Soul.*

ROBERT HAMBOURGER

Assistant Professor at Northwestern University.

RICHARD C. JEFFREY

Professor of Philosophy at Princeton University. Author of *The Logic of Decision, Formal Logic: its Scope and Limits* and (with George Boolos) *Computability and Logic.*

ANTHONY KENNY

Master of Balliol College, Oxford. His publications include *Wittgenstein, Freedom and Power, Descartes* and *The Five Ways.*

NORMAN MALCOLM

Emeritus Professor of Philosophy of Cornell University, Visiting Professor of Philosophy at King's College, London. His publications include *Dreaming, Knowledge and Certainty,* and *Wittgenstein: a Memoir.*

ANSELM WINFRIED MÜLLER

Professor of Philosophy in the University of Trier, West Germany. Author of *Ontologie in Wittgensteins Tractatus.*

HAROLD NOONAN

Lecturer in Philosophy in the University of Birmingham, formerly Research Fellow of Trinity Hall, Cambridge.

CHARLES TAYLOR

Chichele Professor of Social and Political Theory in the University of Oxford, Fellow of All Souls College, Oxford. His publications include *The Explanation of Behaviour* and *Hegel.*

JENNY TEICHMAN

Fellow of New Hall, Cambridge. Author of *The Mind and the Soul* and *The Meaning of Illegitimacy.*

DEIRDRE WILSON

Lecturer in Linguistics at University College, London. Author of *Presupposition and Non-Truth-Conditional Semantics*.

BOGUSŁAW WOLNIEWICZ

Professor of Philosophy in the University of Warsaw. Author of *Rzeczy i Fakty* and translator of major works of Wittgenstein and Frege.

G. H. von WRIGHT

Research Professor in the Academy of Finland, formerly Professor of Philosophy in the University of Cambridge. His publications include *Explanation and Understanding*, *Varieties of Goodness*, *Norm and Action*, and *Time, Change and Contradiction*.

BIOGRAPHICAL NOTE

This collection of essays was presented to G. E. M. Anscombe, Professor of Philosophy in the University of Cambridge, in the year of her 60th birthday. It would be difficult to exaggerate the esteem and affection with which Professor Anscombe is regarded by her friends and colleagues and students: the 17 writers who have contributed to this book represent a very large number of people in Britain, Europe, North America and Australasia who would wish to see her honoured.

Elizabeth Anscombe was born in 1919, the third and youngest child and only daughter of A. W. Anscombe, a schoolmaster at Dulwich College, and his wife Gertrude Elizabeth. Her distinguished, and distinctive, academic career began in 1937 when she went up to Oxford as a Scholar of St Hugh's College. In 1941 she took a First in Greats (Classics and Philosophy). Then she moved to Cambridge where she was for some years a research student at Newnham College. It was in Cambridge, of course, that she met Ludwig Wittgenstein and became his pupil. In 1946 Somerville College, Oxford, appointed her to a Research Fellowship: much of her life's work has been done at Somerville, where she has spent a total of 24 years as Research Fellow and Fellow. She is now an Honorary Fellow of Somerville and also an Honorary Fellow of St Hugh's. In 1967 she was elected Fellow of the British Academy. In 1970 she was appointed to the Chair of Philosophy in the University of Cambridge.

Since Wittgenstein's death in 1951 Elizabeth Anscombe has worked as translator and editor of his posthumous writings, and it is largely to her and to her co-editors, Rush Rhees and G. H. von Wright, that we owe the availability of the Wittgenstein corpus of work—the *Philosophical Investigations*, the *Remarks on the Foundations of Mathematics*, *On Certainty*, *On Colour*, the *Notebooks*, the collections of Wittgenstein's letters and so on. In addition to this work Professor Anscombe has lectured to several generations of students in Oxford, Cambridge and the U.S.A. and has published over 40 articles on Metaphysics, Moral Philosophy, the History of Philosophy, Logic, and Political Philosophy, as well as three books: *An Introduction to Wittgenstein's Tractatus, Intention*, and (with Peter Thomas Geach) *Three Philosophers.* Her versatility is well known but perhaps her most influential writings are those on Causation (the subject of her Inaugural Lecture), on Intention (the topic of her first book), and on Aristotle.

Professor Anscombe has travelled widely and lectured in many countries including the U.S.A., Canada, Poland, Finland, Austria, West Germany, Scotland, Sweden, Spain and Australia. During vacations she sometimes lectures in the University of Pennsylvania as Adjunct Professor. She is married to Professor Peter Thomas Geach and they have three sons and four daughters.

JENNY TEICHMAN
January 1979

PREFACE

> The first statement of the problem of the reality of the past is in Parmenides. "It is the same thing that can be thought and can be", so "what is not and cannot be" cannot be thought. But the past is not and cannot be; therefore it cannot be thought, and it is a delusion that we have such a concept.... Parmenides' remarks suggest the enquiry "How is it that statements about the past have meaning?" (G. E. M. Anscombe, 'The reality of the past')

> Now, empiricist or idealist preconceptions, such as have been most common in philosophy for a long time, are a thorough impediment to the understanding of either Frege or the *Tractatus*. It is best, indeed, if one wants to understand these authors, not to have any philosophical preconceptions at all, but to be capable of being naively struck by such questions as the following ones: If I say that Russell is a clever philosopher, I mention Russell, and say something about him: now, is what I say about him something that I *mention*, just as I *mention* him? If so, what is the connection between these two mentioned things? If not, what account are we to give of the words expressing what I say about him? have they any reference to reality?... The investigations prompted by these questions are more akin to ancient, than to more modern, philosophy. (G. E. M. Anscombe, *An Introduction to Wittgenstein's Tractatus*)

If there is anything that characterizes Elizabeth Anscombe's philosophical writings it is the capacity to be struck by questions. She has made new questions for philosophy: has taken familiar and unquestioned assumptions and shown how far from being obvious they are. Philosophy as she does it is fresh; her arguments take unexpected turns and make unexpected connections, and show always how much there is that had not been seen before. And much is often hard to see, because of our preconceptions. In his contribution to this volume, Richard Jeffrey describes a response which will strike many of her readers as familiar. On first reading a passage in her piece on Aristotle and the sea battle, he says, it struck him as gibberish—but years later its point suddenly became clear to him. She is a leap ahead; but also, characteristically, she can show us how to look two thousand years back to see something which had been lost to sight. She has made the questions of the past live ones for her contemporaries, and sent philosophers back to Hume and Descartes, to Aristotle and Aquinas, Hobbes and Berkeley and Parmenides, whom she has helped us to read in new ways. More than anyone else, she has helped readers of Wittgenstein to find their own way in his writings, and has shown with great imagination and ingenuity the usefulness of his philosophical techniques.

In her inaugural lecture, she called into question the familiar idea of causality as a kind of necessary connection:

> The truth of this conception is hardly debated. It is, indeed, a bit of *Weltanschauung*: it helps to form a cast of mind which is characteristic of our whole culture.

Professor Anscombe has had a go at many bits of *Weltanschauung*. Lucy Brown's contribution to this volume includes an account of what may well have been the first: *The Justice of the Present War Examined.* Written in 1939 while Professor Anscombe was an undergraduate, it expressed strong opposition to the war, and in particular to the demand for unconditional surrender—her position being one which was then extremely unorthodox. Since then, she has attacked many other expressions of the cast of mind characteristic of our whole culture, especially in moral philosophy, and most notably, perhaps, in 'Modern moral philosophy'. What she says, uncosy, unfamiliar, unreassuring, sticks in one's mind—like burrs—and makes one recognize one's parochialisms.

The contributors to this volume of essays, dedicated to Professor Anscombe on her sixtieth birthday, have in common no particular set of views, philosophical or other; what they share is admiration for her and gratitude for her having written and lectured, raised difficulties and sketched solutions, talked and thought and taught. The essays have as their theme intention and intentionality—a pair of subjects to which Professor Anscombe has come back again and again, and whose relation to each other is itself a philosophical matter she has helped to make clear. Besides their concern with a central theme of Professor Anscombe's philosophical writings, these essays bear witness to her sustained interest in the history of philosophy. Her attention to Aristotle, Descartes, Hume and Wittgenstein is reflected in many different ways in the collection. Her second book, *An Introduction to Wittgenstein's Tractatus*, forms the starting point for the paper on

Wittgensteinian semantics contributed by Bogusław Wolniewicz. Five of the papers (by J. E. J. Altham, Roderick Chisholm, Anthony Kenny, Norman Malcolm and Harold Noonan) are responses to her paper 'The first person', which deals with those questions about the self that Descartes put at the centre of philosophical thought. Her treatment of the 'self' as object of thought is continuous with her treatment, nearly a quarter of a century earlier, of past events and past things as objects of thought, in the paper quoted at the beginning of this preface. Both papers show the power of her very characteristic approach, owing much to Wittgenstein, to questions of intentionality. Professor Anscombe once wrote that 'we are now in a position to read Aristotle critically and at the same time with sympathy—without either servility or hostility'. That this is so is in fact partly her own doing—and the papers by Richard Jeffrey, G. H. von Wright and Anselm Müller attest to her success in showing the richness of Aristotle's writings as a field for analytical philosophers. A major theme of Professor Anscombe's has been the criticism of empiricism, in particular empiricism as found in Hume, and the articulation of an alternative position. Her complex response to Hume—hostile respect?—is reflected in the way Humean questions turn up in many of the contributions. For example, Charles Taylor's and Anselm Müller's papers challenge central and familiar theses of Hume's about action, while exploring issues raised by Professor Anscombe in *Intention*. Hume's view of human action is inseparable not only from his general moral theory but also from his views on particular moral issues—as comes out most sharply in what he takes to be the relation between voluntary human action and the course of nature in the essay 'Of suicide'. Professor Anscombe's discussions of war and contraception embody a view strikingly opposed to Hume's, one rooted in an account of intention and action contrasted with his. These papers of hers are discussed—sympathetically and critically—by Lucy Brown and Jenny Teichman. Hume's view of the range of arguments usable to show that the order perceived in the world indicates that it has a designer is criticized in a paper by Robert Hambourger, who makes use of an unpublished paper of Professor Anscombe's on cause and chance.

Peter Geach argues in his paper that there are no logically insulated kinds of truth. It is equally true that there are no logically insulated parts of philosophy—and one argument would itself be the philosophical writings of Professor Anscombe, in which the relations to each other of metaphysics, epistemology, logic and the philosophy of logic, ethics and philosophical psychology can be clearly seen. It is fitting that all of these should be represented in this volume. The four sections into which we have divided the volume indicate some of the main relations between the articles.

The editors are very grateful to all the contributors for their papers, and to Peter Geach for, in addition, much help and good advice. We wish also to give our thanks to Mary Gustafson for permission to reproduce her photograph of Professor Anscombe, and to the University of Virginia for a grant which covered some of the incidental expenses of producing the book.

CORA DIAMOND
January 1979

BIBLIOGRAPHY OF BOOKS AND PAPERS BY G. E. M. ANSCOMBE

Books

Intention, Oxford 1957.

An Introduction to Wittgenstein's Tractatus, London 1959.

Three Philosophers (with Peter Geach), Oxford 1961.

Papers on Metaphysics and the Philosophy of Mind

'Causality and Determination' (inaugural lecture), Cambridge 1971.

'Times, Beginnings and Causes' (British Academy annual philosophical lecture), Oxford 1974.

'Soft determinism' in *Contemporary Aspects of Philosophy* (ed. G. Ryle), London 1977.

'Causality and extensionality', *Journal of Philosophy* vol. LXVI no. 6, 1969.

'Before and after', *Philosophical Review*, 1964.

'The reality of the past' in *Philosophical Analysis* (ed. Max Black), N.J. 1950.

'Memory, experience and causation' in *Contemporary British Philosophy* (ed. H. D. Lewis) fourth series, London 1974.

'Substance', *Proceedings of the Aristotelian Society* supplementary volume XXXVIII, 1964.

'Intention', *Proceedings of the Aristotelian Society*, 1957.

'Pretending', *Proceedings of the Aristotelian Society* supplementary volume XXXII, 1958.

'On the grammar of "enjoy"', *Journal of Philosophy* vol. LXIV no. 19, 1967.

'On sensations of position', *Analysis* vol. XXII no. 3, 1962.

'The intentionality of sensation' in *Analytical Philosophy* (ed. R. J. Butler) second series, Oxford 1965.

'The subjectivity of sensation', AJATUS 36 (yearbook of the Philosophical Society of Finland 1976).

'The first person' in *Mind and Language* (ed. Samuel Guttenplan), Oxford 1975.

'Subjunctive conditionals', *Ruch Filozoficzny* vol. XXXIII no. 3–4, Poland 1975.

'Under a description', *Nous*, 1979.
Analysis Competition number 10: set and judged by G. E. M. Anscombe, won by 'Al Tajtelbaum', *Analysis* vol. XVI no. 6, 1956, vol. XVII no. 3, 1957.
'A reply to Mr. C. S. Lewis's argument that "naturalism" is self-refuting', *Socratic Digest*, Oxford *circa* 1947.

Papers on the History of Philosophy

'Parmenides, mystery and contradiction', *Proceedings of the Aristotelian Society*, 1969.
'The new theory of Forms', *The Monist* vol. L no. 3, 1966.
(There is an unpublished companion piece entitled 'The early theory of Forms'.)
'Understanding Proofs', *Philosophy* vol. LIV no. 208, 1979.
'Aristotle and the sea battle', *Mind* vol. LXV no. 257, 1956.
'The principle of individuation' (on Aristotle), *Proceedings of the Aristotelian Society* supplementary volume XXVII, 1953.
'Thought and action in Aristotle' in *New Essays on Plato and Aristotle* (ed. J. R. Bambrough), London 1965.
'Necessity and truth' (on Aquinas), *Times Literary Supplement* 14 February 1965.
'Hume and Julius Caesar', *Analysis* vol. XXXIII no. 1, 1973.
'"Whatever has a beginning of existence must have a cause": Hume's argument exposed', *Analysis* vol. XXXIV no. 1, 1974.
'Will and emotion' (on Brentano), *Gräzer Philosophische Studien* vol. 5 (ed. Rudolf Haller) Graz, Austria 1978.
'Retractation' (on Wittgenstein), *Analysis* vol. XXVI no. 2, 1966.
'The question of linguistic idealism' in *Essays on Wittgenstein in Honour of G. H. von Wright* (ed. J. Hintikka), Holland 1976.

Papers on Ethics, Politics, and the Philosophy of Religion

'Does Oxford Moral Philosophy corrupt the youth?', *Listener* 14 February 1957.
'Modern moral philosophy', *Philosophy* vol. XXXIII no. 124, 1958.
'On brute facts', *Analysis* vol. XVIII no. 3, 1958.

'Two kinds of error in action', *Journal of Philosophy* vol. LX no. 14, 1963.
'On promising', *Critica* vol. III no. 7/8, Mexico 1969.
'The Justice of the Present War Examined' (with Norman Daniel). Published by the authors, Oxford 1939.
'Mr Truman's Degree.' Published by the author, Oxford 1957.
'War and murder' in *Nuclear Weapons: A Catholic Response* (ed. Walter Stein), London and New York 1961.
'You can have sex without children: Christianity and the new offer' in *Renewal of Religious Structures: Proceedings of the Canadian Centenary Theological Congress*, Toronto 1968.
'Contraception and chastity', *The Human World,* May 1972.
Reply to Peter Winch, B. A. O. Williams and M. K. Tanner, *The Human World*, November 1972.
'Contraception and Chastity', Catholic Truth Society, London 1977.
'Rules, rights, and promises', *Mid-Western Studies in Philosophy* (iii), 1978.
'On frustration of the majority by fulfilment of the majority's will', *Analysis* vol. XXXVI no. 4, 1976.
'On the source of the authority of the State', *Ratio*, 1978.
'On Transubstantiation', Catholic Truth Society, London 1974.

Translations

Ludwig Wittgenstein: *Philosophical Investigations*, Oxford 1953.
Remarks on the Foundations of Mathematics, Oxford 1956.
Notebooks 1914–1916, Oxford 1961.
Zettel, Oxford 1967.
On Certainty (with Denis Paul), Oxford 1969.

René Descartes: *Philosophical Writings*—a selection including the *Meditations, Objections and Replies (the third set),* the *Discourse on Method, Principles of Philosophy*, the *Dioptrics*, a selection of letters, etc. (with P. T. Geach), London 1954.

I

THE FIRST PERSON

[1] THE FIRST PERSON
Anthony Kenny

As a graduate student at St Benet's Hall, Oxford, from 1957–1959 I attended the classes on Wittgenstein given by Elizabeth Anscombe in a chilly and dilapidated outhouse of Somerville College. I look back on those classes as the most exciting and significant event in my education in philosophy. Like many others, when I came to the classes I regarded Wittgenstein's attack on private languages with incomprehension mixed with hostility. Miss Anscombe encouraged us to give the fullest possible expression to our doubts and disagreements: from time to time I found myself thrust into the uncomfortable position of spokesman for the pro-private-language party. By the end of the term I had become convinced of the correctness, and the profound importance, of the insights expressed by Wittgenstein in his critique of private ostensive definition. The seminar completely changed the way in which I looked at issues in philosophy of language and philosophy of mind: various lines of thought which until that time I had found seductive, and which many others still follow enthusiastically, lost all their attraction and were revealed as blind alleys and dead ends. The lines of thought ramified over all areas of philosophy, but all of them can broadly be termed Cartesian.

One thing which I learned from reading Wittgenstein with Miss Anscombe was to have an enormous respect for the genius of Descartes. Those who accept a Cartesian view of the mind, I suppose, can admire Descartes for being the first to state certain truths with cogency and elegance and concision. But only one who is cured of Cartesianism can fully be awed by the breathtaking power of an intellect which could propagate, almost unaided, a myth which to this day has such a comprehensive grasp on the imagination of a large part of the human race. To those who doubt the power of Cartesian ideas to survive and flourish in the most hostile of climates, I commend a reading of Professor Anscombe's paper 'The first person'.[1]

Professor Anscombe's paper takes its start from Descartes' argument to prove that he is not a body. The argument, she observes, is essentially a first-person one, which each of us must go through for himself. The conclusion of the Cartesian doubt could as well have been 'I am not Descartes' as 'I am not a body': Descartes might have concluded to the non-identity of himself with Descartes. This is not the self-contradiction that it seems, because the 'himself' here is not a pronoun replaceable by 'Descartes': it is the indirect reflexive, which has to be explained in terms of 'I'.

We cannot explain 'I' by saying that it is the word each of us uses to refer to himself. If the 'himself' here is the ordinary reflexive pronoun, then the specification is inadequate: one can refer to oneself without knowing one is doing so, and in general knowledge of the referent of a referring expression does not amount to knowledge of its sense. If the 'himself' is the indirect reflexive then the account is circular since the indirect reflexive is simply the *oratio obliqua* version of the *oratio recta* 'I'.

'I' is not a proper name. This is not because it is a name which everyone has: that would be perfectly conceivable. Nor is it because it is a name that everyone uses only to speak of himself. Such a situation too would be imaginable: suppose everyone had '*A*' marked on his wrist, and one of the letters '*B*' to '*Z*' marked on his back; it might be that everyone reported on himself by using '*A*' and on others by using '*B*' to '*Z*'—that would still not make '*A*' like our 'I' unless it was, like 'I', a manifestation of self-consciousness.

What is self-consciousness? Is it a consciousness of a self, a self being something that some things have or are? If so, then we might conceive 'I' as the name of a self, and an account of what kind of thing a self was would clarify the use of 'I' in the way that an account of what a city is could communicate part of what is needed to understand the use of a name like 'London'. But self-consciousness, Professor Anscombe argues, is not consciousness of a self; it is simply consciousness that such-and-such holds of oneself, where 'oneself' is again the indirect reflexive. The very notion of a self is begotten of a misconstrual of this pronoun.

If 'I' is not a proper name shall we simply say then that it is a pronoun? The grammatical category of pronouns is a ragbag, including even variables; and the suggestion given by the word's etymology, that it can be replaced by a noun in a sentence while preserving the sense of the sentence, is false of 'I'. Shall we say that 'I' is a demonstrative? If we use a demonstrative like '*this*' we must be prepared to answer the question 'this *what*?'; it is not clear what the corresponding question and answer is with 'I'. Moreover, a demonstrative like 'this' may fail to have a reference (for example if I say 'this parcel of ashes' pointing to what, unknown to me, is an empty box); no such failure of reference is possible, it seems, if we use 'I' to refer.

In fact, Professor Anscombe insists, 'I' is neither a name nor any other kind of expression whose logical role is to make a reference at all. Of course, it is true that if *X* makes assertions with 'I' as grammatical subject, then those assertions will be true if and only if what he asserts is true of *X*. But this doesn't mean that 'I' refers to *X*, for the truth-condition of the whole sentence does not determine the meaning of the items within the sentence. One who hears or reads a statement with 'I' as subject needs to know whose statement it is if he wants to know how to verify it. But that does not make 'I' a referring expression, any more than the '– o' at the end of a Latin verb such as '*ambulo*', which signifies the same requirement.

If 'I' were a referring expression at all, it would seem to be one whose reference is guaranteed in the sense that the object an 'I'-user means by it must exist so long as he is using 'I', and in the sense that he cannot take the wrong object to be the object he means by 'I'. The only thing thus guaranteed is indeed the Cartesian Ego: certainly not the body.

> Imagine that I get into a state of 'sensory deprivation'. Sight is cut off, and I am locally anaesthetized everywhere, perhaps floated in a tank of tepid water; I am unable to speak, or to touch any part of my body with any other. Now I tell myself "I won't let this happen again!" If the object meant by "I" is this body, this human being, then in these circumstances it won't be present to my senses; and how else can it be 'present to' me? But have I lost what I mean by "I"? Is that not present to me? Am I reduced to, as it were, 'referring in absence'? I have not lost my 'self-consciousness'; nor can what I mean by "I" be an object no longer present to me.

Thus if 'I' is a referring expression, Descartes was right about its referent: though his position runs into intolerable difficulties about the reidentification of the Ego from thought to thought.

The only way to avoid the Cartesian blind alley, Professor Anscombe maintains, is to break altogether with the idea that 'I' refers at all. If we give up this idea, we must also recognize that 'I am N.N.' is not an identity proposition. 'N.N. is this thing here'—'N.N. is this body'—'N.N. is this living human being': these are all identity propositions. But to get from them to 'I am N.N.' we need the proposition 'I am this thing here'—and this is not an identity proposition.

The kernel of Anscombe's positive account is given in these paragraphs:

> "I am this thing here" is, then, a real proposition, but not a proposition of identity. It means: this thing here is the thing, the person (in the 'offences against the person' sense) of whose action *this* idea of action is an idea, of whose movements *these* ideas of movement are ideas, of whose posture *this* idea of posture is the idea. And also, of which *these* intended actions, if carried out, will be the actions. . . .
>
> If I were in that condition of 'sensory deprivation', I could not have the thought "this object", "this body"—there would be nothing for "this" to latch on to. But that is not to say I could not still have the ideas of actions, motion, etc. For these ideas are not extracts from sensory observation. If I do have them under sensory deprivation, I shall perhaps *believe* that there is such a body. But the possibility will perhaps strike me that there is none. That is, the possibility that there is then nothing that I am.
>
> If "I" were a name, it would have to be a name for something with this sort of connection with this body, not an extra-ordinary name for this body. Not a name for this body because sensory deprivation and even loss of consciousness of posture, etc., is not loss of *I*. (That, at least, is how one would have to put it, treating "I" as a name.)
>
> But "I" is not a name: these I-thoughts are examples of reflective consciousness of states, actions, motions, etc., not of an object I mean by "I", but of this body. These I-thoughts (allow me to pause and think some!) . . . are unmediated conceptions (knowledge or belief, true or false) of states, motions, etc., of this object here, about which I can find out (if I don't know it) that it is E.A. About which I did learn that it is a human being.

Most people with whom I have discussed this article find its destructive arguments unconvincing and its negative conclusion preposterous. For myself, I am wholly persuaded that 'I' is not a referring expression, and that 'I am N.N.' is not an identity proposition. I accept that 'the self' is a piece of philosophers' nonsense produced by misunderstanding of the reflexive pronoun—to ask what kind of substance my *self* is is like asking what the characteristic of *ownness* is which my own property has in addition to its being *mine*. I accept that a person is a living human being, and that I am such a person and not a Cartesian ego, a Lockean self, or an Aristotelian soul. None the less, I find Anscombe's positive account of 'I' unacceptable. Astonishingly, it seems to me, she falls into the Cartesian trap from which Wittgenstein showed us the way out.

Consider the sentence 'This body is the person of whose action *this* idea of action is an idea'. What is *this* idea of action? As Professor Anscombe uttered these words in her lecture, perhaps she had a mental image of herself waving an arm, or had the thought 'I will wave my arm'. This kind of thing, no doubt, is what she was referring to by the expression 'idea of action'. But what is the role of the demonstrative '*this*'? It was not meant to single out one idea of action from among other items in her mental history: it was not meant to contrast, say, the idea of waving an arm with the idea of putting the left foot forward. 'This', in her mouth, in that context, was simply tantamount to 'my'. It was not, of course, an invitation to her hearers to inspect her mental images or to fix their attention on her secret thoughts. But since it was not that, could the remark give her hearers any information at all? To say 'My body is the body of whose action my idea of action is an idea' is not to say anything that could possibly be false; and 'this body is my body' is equally truistic if 'this body' means 'the body uttering this sentence'. We individuate people's ideas of action by individuating the bodies that give them expression. When we in the audience listened to Professor Anscombe's sentence we did not first locate the idea of action, and then identify the body meant by 'this body' and finally grasp the relation intended between the two. We were inclined to assent to what was said simply because the same body, the same person, was speaking throughout the sentence.

But did not Professor Anscombe make clear *to herself* what she meant by '*this* idea of action' and do so independently of the truth of the 'real proposition' which she expressed by saying 'I am this thing here'? Could she then have been in doubt of, or ignorant of, *which* idea of action was meant, and did the mental attention accompanying 'this' remove or prevent the doubt? It is here that Wittgenstein's critique of the notion of private ostensive definition becomes relevant. '*This* idea of action' is not, of course, the expression of an attempt at private ostensive *definition*: it is not meant to give *sense* to the expression 'idea of action', but only to indicate its reference. And Wittgenstein's target was the idea of private sense, not of private reference; nothing in what he says rules out the possibility of referring to our own and others' secret thoughts. But if Wittgenstein is right, any private ostension or demonstration must be something which it would make sense to think of as being done publicly: one can refer, for instance, to the content of a dream or mental image because one could exhibit it by narrating or drawing it. But when the '*this*' in '*this* idea of action' is meant to mean 'mine, not someone else's', there does not seem to be anything that could be a public performance of this private demonstration without being at the same time a pointing to *this body*. Hence it seems that '*this* body is the body *these* ideas are about' cannot be a 'genuine proposition' if that means a proposition that could be used to convey information.

But even in the public case, it may be said, the body that gives expression to an idea of action need not be the body that enters into the content of the idea. When *X* says to *Z* through an interpreter *Y* 'I will meet you at the airport at 10.30' it is *Y*'s body that produces the sounds, *X*'s body that verifies or falsifies the pronouncement. But this does not drive a wedge between the individuation of the idea of action and the individuation of its content: it merely shows that the notion of 'expression' is not a simple one. In such a case, the idea of action is not *Y*'s idea any more than its fulfilment is: when we look for the verifier or falsifier of 'I'-sentences we have to look for the primary utterer of the sentence and not secondary utterers such as interpreters, telephones, or tape-recorders. We individuate the idea of action by individuating the primary utterer; and we discover that *X* is the primary utterer in such a case by discovering, *inter alia*, the sounds made by *X*'s body within the range of detection of the secondary utterer. If an interpreter relayed Anscombe's lecture to a foreign audience, the expression 'this idea of action' and the expression 'this body' would refer to Anscombe's idea and Anscombe's body, not to the interpreter's.

But are there not cases where speakers disown the sounds coming out of their mouths, without there being any other person or body which can be identified as the primary utterer? Professor Anscombe reminds us of the utterances of mediums and of those who believe themselves possessed by spirits. She invites us to consider the situation in which someone stands before us and says 'Try to believe this: when I say "I", that does not mean this human being who is making the noise. I am someone else who has borrowed this human being to speak through him'. Does not a consideration of such cases show that 'I am this thing here' *is* a genuine proposition, which could serve to convey the information that the listener is *not* face to face with a case of possession or communication through a medium?

We should observe first that the possibility of 'I'-utterances being given a verification by something other than the utterer does not by itself suffice to give content to the notion of control by spirits. Suppose that a sybil in a trance says 'I will destroy Jerusalem on Christmas Day 1984'; and suppose that on Christmas Day 1984 Jerusalem is ravaged by an earthquake or sacked by an army out of control. This, by itself, does not enable us to answer, or even coherently to raise, the question 'And who was the "I" who made the prediction'? Consider the clause in Anscombe's imagined utterance 'When I say "I" . . .'. We can only make something of this because we tacitly accept the first 'I' as standing at least temporarily in the normal relation to the human being who utters it: if we set ourselves to obey the spirit's instructions, it will be the next 'I's uttered *by the same human being* that we will respectfully attend to.

But perhaps Professor Anscombe's logical point can be made without introducing the difficult notion of spiritual agency. Imagine two Siamese twins related in the following way: whenever the mouth of Tweedledum says 'I will do *X*', the body of Tweedledee does *X*, and vice versa; and whenever something happens to the body of Tweedledee it is reported in the first person by Tweedledum. Do we not now have a case where the body which owns an 'I'-idea is distinct from the body which provides its subject matter? Cannot either twin say 'I am not *this* body, I am *that* body'? And so have we not a conceivable situation which might be ruled out by the 'genuine proposition' 'I am this body here'?

Nothing in the situation, it seems, *compels* us to say that the body which utters 'I' is not the body which the 'I'-utterances are about. No doubt there is little to attract us in the suggestion that perhaps the 'I'-utterances are about the uttering body, but are just all false—though the body which the 'I'-utterances are about is of course the one which verifies *or falsifies* them. But might not one say that what we have here is a pair of bodies of unusual shapes, with the mouth, instead of being central, being, as it were, an offshore island?

But even if we do say that Tweedledum's mouth utters 'I'-sentences which are about Tweedledee's body we cannot, it seems, take these 'I'-sentences to be expressions of self-consciousness in the way that normal 'I'-sentences are. For let us suppose that Tweedledum has a thought about himself which he wishes to communicate to others: suppose, for instance, he has the thought 'I will tell Professor Anscombe about my pitiable condition'. How is he to carry out this resolve? Which mouth will he use to express the thought 'I am not this body'? If he uses Tweedledee's mouth, then that mouth will say 'I am not this body'. But *that*, if it communicates information at all, communicates information about Tweedledee, not about Tweedledum. If Tweedledum uses his own mouth to express the information, then it is not the case that the 'I'-ideas of Tweedledum are always ideas of the action of Tweedledee's body. We can preserve our fiction from collapsing only if we forget that the expression of a thought by uttering it is as much the enactment of an idea as any other bodily movement. If we exclude the expressive function of 'I'-utterances, then the 'I'-utterances of the twins resemble rather the '*A*'-utterances imagined by Professor Anscombe than the 'I'-utterances of self-conscious persons.

Such fantasies, and the phenomena of possession, may throw into confusion the sense of 'I' emerging from a human mouth. They cannot create a sense of 'I' in which, instead of expressing the self-consciousness of the body to which the mouth belongs, it serves to express the self-consciousness of some other agent. Someone believing himself possessed may of course say 'It was not I who uttered those terrible blasphemies a moment ago'. He is not denying that it was his body that made the noises: and we can make sense of his claim. Not every movement that a person's body makes is a bodily movement of that person: but that does not mean that there can be bodily movements of a person which are not movements of that person's body. For a movement of *X*'s body to be a movement made by *X* it must be a voluntary movement: and the blasphemies of the possessed, we may believe, are not voluntary actions of theirs. But in all this it is the 'I' of the unfortunate person possessed that we can understand, not the 'I' of a possessor.

Let us turn from possession to sensory deprivation. Does the consideration of this enable us to sever the 'I'-thoughts from the body which verifies or falsifies them? Professor Anscombe seems to suggest that in a state of sensory deprivation I can think, and privately concentrate on, 'I'-thoughts (for example 'I won't let this happen again') without knowing which body they concern, and indeed without knowing whether I have a body at all. But if these thoughts have a sense, the sense must surely be expressible; and what they say could only be said publicly if it were said by the right body. Indeed, if I can be said to be in doubt whether I have a body, then the sense of 'I won't let this happen again' must surely be in question. Such a resolve, or plan of action, is a plan *for a body*. The resolve is not like 'I will drive my car home now' said when I do not know that my car has been stolen. 'I will drive my car home, if I still have a car' makes sense; but 'I will get out of this bath, if I still have a body' does not. For if I no longer have a body, then I no longer exist, as Professor Anscombe explicitly concedes. And if I do not exist, then I cannot be making resolves either. Following the path of Cartesian doubt seems to lead to a very unCartesian thought: 'Perhaps I don't exist, but if I do exist, I'll never let this happen again'.

Thoughts may be kept to oneself; but even the most secret thought must be capable of being made public, and the sense of the thought expressed in public must be the same as the sense of the thought entertained in private—otherwise we could not speak of the 'expression' of the thought. Professor Anscombe has argued convincingly that 'I' is not a word whose function is to *refer* to its utterer. But it is part of the *sense* of 'I' that assertions containing it are verified or falsified by reference to the history of the assertor. That is something that you know by knowing the grammar of the word, not something you know as a result of study of the context of a particular utterance. A public utterance of an 'I'-sentence has a clear sense only when it is clear which is the body, the person, that utters it as its primary utterer. What goes for the public utterance must go for the private thought too, if Wittgenstein is right that there is no such thing as private sense.

Of course, in a normal case, where there is no sensory deprivation, there may be genuine questions of the form 'Is this body my body?' I may see a body in a mirror, or glimpse part of a trouser leg under the table, for instance, and raise the question. But when these questions are answered, the uncertainty that is removed is an uncertainty about which body *this* is, not an uncertainty about which body *mine* is. In the case of sensory deprivation I cannot have thoughts of the form 'this body is my body', for there is nothing for the 'this' to latch on to, no glimpse or sound or sensation. But what sensory deprivation cuts me off from is the reference of 'this', not the sense of 'my'. It is not as if there is an enterprise of identifiying my own body, which I can do in the normal case and fail to do in the case of sensory deprivation. In calling a body 'my body' I do not identify the body, either for myself or for anyone else; though, for others, I may thereby identify *myself.*

In the normal case, it is not by sensory experience that I know I have a body; the lack of sensory experience therefore does not prevent me knowing I have a body, and does not prevent my 'I'-thoughts from being about that body. If it did render it uncertain whether I had a body it would, for the reasons given, render unclear the sense of the 'I' in the 'I'-thoughts. Once again, the attempt to provide a contrast to give content to the utterance 'I am this body' fails. 'I am this body' can be given a sense in particular circumstances, as when pointing to a photograph or a mirror; it does not have a general sense as expressing a truth which each of us knows about himself but cannot communicate to others.

The conclusion of Professor Anscombe's article seemed to

be this. 'I am the thinker of these thoughts' is not a genuine proposition, but 'I am this body' is a real proposition which answers a real question. This conclusion I found surprisingly unWittgensteinian: the thinker of these thoughts who is possibly not this person with this body seems uncomfortably similar to a Cartesian ego; and if we allow the conceivability of the notion that we are thus spiritual, it seems no great matter that when we use 'I' we are not actually *referring* to any such spirit or self. I have argued that neither consideration of the role of interpreters, nor reflection on the phenomena of possession, nor imagination of a state of sensory deprivation give reason for thinking that content can be given to 'I'-thoughts where there is no person identifiable as an actual or possible utterer of the thoughts. We cannot drive a wedge between the body that expresses a first-person idea of action, and the body that is the subject-matter of the idea of action, because it is part of the sense of 'I' that utterer and subject should be one and the same. Any circumstances which we could imagine which would suggest a divorce between the two would to the same extent call in question the sense of the 'I' in the 'I'-thoughts.

The arguments I have used in this paper are all derived from principles made familiar by Wittgenstein's critique of private ostensive definition: principles I first came to appreciate at those classes in Somerville some twenty years ago. It may be that I have misunderstood Professor Anscombe's article: I am not sure that I have understood it correctly, but perhaps my expression of my interpretation of it may help others to see it in a clearer light. Whether or not the objections which I have made to it stand up, the article seems to me to be of great interest—an interest which is independent of the question whether it marks the conversion of Professor Anscombe from Wittgenstein to Descartes.

Note

1 G. E. M. Anscombe, 'The first person', in *Mind and Language*, ed. S. Guttenplan (Clarendon Press, Oxford, 1975), pp. 45–66.

[2] WHETHER 'I' IS A REFERRING EXPRESSION

Norman Malcolm

I wish to discuss Professor Elizabeth Anscombe's powerful paper, 'The first person'.[1] Her principal theme is that the first person pronoun 'I' is not a referring expression. To be sure, there is an overwhelming inclination to suppose that when a speaker utters a sentence in which 'I' occurs as grammatical subject, then *of course* the speaker uses 'I' to refer to, to stand for, to designate, one particular individual thing; one is tempted to suppose that the very function of the first person pronoun in any language that has it, is to enable a speaker to make a unique reference, to select out one and only one thing, from all the beings, things, or objects that there are in the world. Descartes certainly believed that 'je', 'moi', 'ego', were used by him to refer to a certain thing. Having apparently discovered that *je pense, donc je suis* is a truth that cannot be shaken by even the most extravagant scepticism, Descartes then proceeded, as he reports in Part IV of the *Discourse*, to examine with attention 'ce que j'étais'—in other words, to try to find out what is the nature of that thing to which he refers by 'je' or 'moi'. His conclusion was that 'j' étais une substance dont toute l'essence ou la nature n'est que de penser'. He is supposing that he uses 'je' to refer to a 'substance'; and by a 'substance' he certainly means a *thing*, that has a nature and properties. He draws the further conclusion that 'ce moi' is entirely distinct from the body, and easier to know than the body, and that even if there was no body 'ce moi' would still be just what it is. Similarly, in the *Second Meditation*, when he asks 'But what then am I? (*Sed quid igitur sum*?), he surely seems to be asking 'What sort of thing is this 'I'-thing?'. His answer is 'A thinking thing' (*Res cogitans*).

Hume speaks rather scornfully of those philosophers

> . . . who imagine that we are every moment intimately conscious of what we call our SELF; that we feel its existence and its continuance in existence; and are certain, beyond the evidence of a demonstration, both of its perfect identity and simplicity.[2]

Hume declares that he is not aware of any such thing:

> For my part, when I enter most intimately into what I call *myself*, I always stumble on some particular perception or other, of heat or cold, light or shade, love or hatred, pain or pleasure. I never can catch *myself* at any time without a perception, and never can observe any thing but the perception.

Hume still assumes, however, that 'I' and 'myself' are referring expressions; they refer, however, not to an enduring thing of 'perfect identity and simplicity', but to a collection or 'bundle' of perceptions. In an Appendix to the *Treatise* he says:

> When I turn my reflexion on *myself*, I never can perceive this *self* without some one or more perceptions; nor can I ever perceive any thing but the perceptions. 'Tis the composition of these, therefore, which forms the self.

Yet Hume confesses himself to be baffled by this view, because he cannot 'explain the principle of connexion' that unites the successive perceptions.

Thomas Reid rejects Hume's proposal that the *I* or *self* is a collection of successive thoughts, feelings, sensations, and reaffirms the Cartesian view:

> I see evidently that identity supposes an uninterrupted continuance of existence. That which hath ceased to exist, cannot be the same with that which afterwards begins to exist; for this would be to suppose a being to exist after it ceased to exist, and have had existence before it was produced, which are manifest contradictions. Continued uninterrupted existence is therefore implied in identity.[3]

Reid goes on to say:

> My personal identity, therefore, implies the continued existence of that indivisible thing which I call myself. Whatever this self may be, it is something which thinks, and deliberates, and resolves, and acts, and suffers. I am not thought, I am not action, I am not feeling; I am something that thinks, and acts, and suffers. My thoughts and actions, and feelings, change every moment—they have no continued, but a successive existence; but that *self* or *I*, to which they belong, is permanent, and has the same relation to all the succeeding thoughts, actions, and feelings, which I call mine.

Reid seems to suppose that there is an individual thing which *he* designates by the words 'I' and 'myself'. In his remark that his personal identity 'implies the continued existence of that indivisible thing *which I call myself*' (emphasis added), there is a strong hint that 'I' and 'myself' are used by him as *names* of an entity that endures continuously throughout Reid's life. Reid implies of course that every English speaker uses 'I' and 'myself' as Reid does; that is to say, each speaker uses those pronouns to designate an entity that exists continuously throughout the life of the speaker. But of course each speaker uses those pronouns to refer to an entity that is numerically different from the entity that any other speaker refers to when he employs those pronouns.

Thus Reid appears to have held the view which Anscombe characterizes as the view that not only is 'I' a name, but that it is 'logically a proper name'.[4] One feature of the philosophical notion of that which is 'logically a proper name' is that it has, as Anscombe puts it, a 'guaranteed reference'. Russell said of the name 'Romulus' that it is not really a name, but a truncated or telescoped description: 'If it were really a name, the question of existence could not arise, because a name has got to name something or it is not a name'.[5] The 'names' of the *Tractatus* had this feature of guaranteed reference because the object that a *Tractatus* name meant or stood for was *the meaning* of the name.[6] Anscombe remarks that *if* 'I' is a name, that is, if its job is to refer to or designate something, then it has this feature of guaranteed reference:

> "I"—if it makes a reference, if that is, its mode of meaning is that it is supposed to make a reference—is secure against reference-failure. Just thinking "I. . . ." guarantees not only the existence but the presence of its referent. It guarantees the existence *because* it guarantees the presence, which is presence to consciousness. But N.B. here "presence to consciousness" means physical or real presence. For if the thinking did not guarantee the presence, the existence of the referent could be doubted. For the same reason, if "I" is a name it cannot be an empty name. I's existence is existence in the thinking of the thought expressed by "I. . . ." This of course is the point of the *cogito*. . . . (pp. 54–55)

But if 'I' is a name, or if not a name then perhaps a demonstrative, what is the nature of the object to which it refers? There have been several different ideas about this. First, there is the notion of Descartes and Reid that the object to which 'I' refers is something that is indivisible, immaterial, containing no parts; is that which thinks, acts, suffers; and is permanent throughout the life of a human being. In short, it is a Self or Soul. A second idea is that the object to which 'I' refers is a living, walking, speaking, human being, such as N.M. or E.A. Obviously these are different ideas about the reference of 'I'. The living, walking, speaking N.M. is not indivisible; for N.M. could be drawn and quartered. A third notion is that 'I' refers to a sequence or 'bundle' of thoughts, sensations, feelings. Anscombe proposes a fourth candidate, namely, that what 'I' refers to is *this body* (pp. 57–58).

Let us consider these candidates in reverse order. Anscombe calls attention to the possibility of one's getting into a state of 'sensory deprivation', in which one could not see, hear, talk, or feel anything by touch, or have any sensations in or on one's body, because all of it is anaesthetized. If I were in a condition of sensory deprivation, *this body* would not be present to my senses; moreover, there is no way at all in which it would be *present to me* (p. 58). The postulated condition of sensory deprivation need not, however, be a condition of unconsciousness. I could think to myself: 'I am in a strange condition'. I might even wonder whether I had become disembodied. I could have many other 'I'-thoughts. But as long as I was in the state of sensory deprivation, the 'I' in those 'I'-thoughts would not designate or denote *this body*. Thus if 'I' is supposed to be a referring expression with a guaranteed reference to some one and the same thing, then *this body* does not fill the bill.

The proposal that 'I' refers to a course, flow, sequence, stream, collection, group, bundle, or totality, of sensations, thoughts and feelings, seems quite worthless. If 'I' is supposed to refer to each different thought or sensation as it occurs, then the reference of 'I' would be continually shifting, often with great rapidity. It would be hard to see the difference between the word 'I' having that sort of reference and its not being a referring expression at all. On the other hand, if 'I' is not supposed to refer to the various particular thoughts, sensations, feelings, but instead to that by virtue of which they are the thoughts, etc., of some one and the same person, then it is impossible to understand, as Hume realized, what the connecting principle might be; and also impossible to see what role there would be for 'I' as a referring expression. If we say that the object that 'I' designates is a thing that *has* those thoughts then, of course, we have moved entirely away from Hume, and towards one or the other of the first two theories.

Let us turn to the second view mentioned in our list, namely, the view that 'I' is used by each speaker to refer to a living, corporeal, human being, with a particular history, such as N.M. or E.A. It is true that on this view there will be, in a sense, a guaranteed reference for 'I'. If *I* say, 'I broke the vase', this is true if and only if N.M. broke the vase; and if *I* say 'I am confused' this is true if and only if N.M. is confused. Every speaker or thinker is a real person, X, and every 'I'-thought will be a thought of the person, X, who is the thinker. But, as Anscombe points out, if someone

> . . . takes this as an adequate account of the guaranteed reference of "I", then he will have to grant that there is a further sort of 'guaranteed reference', which "I" does *not* have. Guaranteed reference for that name "X" in this further sense . . . would entail a guarantee, not just that there is such a thing as X, but also that what I take to be X *is* X. (p. 57)

It is not true, however, that when I use 'I' it has guaranteed reference in this further sense. Through being misinformed or duped, or from some mental aberration, I might believe I was John Smith, not merely bearing the name 'John Smith' but also having the origin and history of John Smith. Yet if I said 'I am writing a letter', this could be true even though I mistakenly believed that I was John Smith. My mistaken identification of myself as John Smith would have no bearing at all on the truth or falsity of my statement, 'I am writing a letter'. This proves that the 'I' in that statement cannot *mean* or *stand for* John Smith, since if it did the statement would be false. Anscombe says:

> It seems clear that if "I" is a 'referring expression' at all, it has both kinds of guaranteed reference. The object an "I"-user means by it must exist so long as he is using "I", nor can he take the wrong object to be the object he means by "I". (p. 57)

This is an oblique way of arguing that there is *no* object the 'I'-user means by 'I'. If he mistakenly believed himself to be such-and-such a person, and if he did use 'I' to mean that person, then he would be taking the wrong object to be the object he means by 'I'.

It is obvious, however, that if I had false beliefs about my origin and course of life, this would be irrelevant to the truth and the meaning of such statements of mine as that 'I am tired', 'I am seated in a chair', 'I intend to go to the movies'. This would also be the case if, suffering from amnesia, I was totally bewildered about my identity—had no idea at all as to who I was.

Wittgenstein remarks that when I say 'I'm in pain', I am *not* saying that such-and-such a person is in pain.[7] And in *The Blue Book* he makes the seemingly paradoxical remark that 'In "I have pain", "I" is not a demonstrative pronoun.'[8] The impression of paradox will disappear if we ask ourselves, 'What job does a demonstrative do?' The answer is that the words 'this', 'that', 'he', 'she', 'it', are used to pick one thing out from others. Which person or thing the demonstrative word picks out is shown by a pointing gesture, or by a previously given name or description, or by the context of previous or subsequent remarks. Wittgenstein's point is that 'I' is not used like that. It does not mean 'such-and-such a person'. It does not designate 'a particular person'.[9] My saying, 'I'm in pain', may of course attract the attention of others to myself; but so also may my silent grimacing with pain. I do not, in the first case any more than in the second, mention, refer to, or mean one particular person among others.

Of course by 'I' I mean *myself.* But this is equivalent to: by 'I' I mean *I*; which says nothing, being a kind of tautology. If my 'I' meant a person of a particular identity, origin, and course of life, and I was not that person, then my statement, 'I am standing up', would be false, even if I was standing up—which is absurd.

This brings us to the first candidate on our list, namely, the notion that 'I' refers to the Cartesian Ego, Self, or Soul—the thinking thing. This seems to be the only possibility. 'Thus we discover that *if* "I" is a referring expression, then Descartes was right about what the referent was' (p. 58). But this view of Descartes, and of Reid, has nonsensical consequences. As Anscombe points out, not only is there 'the intolerable difficulty of requiring an identification of the same referent in different "I"-thoughts', but also there could be no guarantee that there was only *one* thinker of a single 'I' thought: 'How do I know that "I" is not ten thinkers thinking in unison'? (p. 58).

Anscombe gives an accurate summary of the matter:

> Getting hold of the wrong object *is* excluded, and that makes us think that getting hold of the right object is guaranteed. But the reason is that there is no getting hold of an object at all. With names, or denoting expressions (in Russell's sense) there are two things to grasp: the kind of use, and what to apply them to from time to time. With "I" there is only the use. (p. 59)

The solution of the problems about the reference of 'I' is this: '"I" is neither a name nor another kind of expression whose logical role is to make a reference, *at all*' (p. 60).

In reflecting on Anscombe's paper I have wondered whether 'I' might sometimes be a referring expression and sometimes not. In *The Blue Book* Wittgenstein seems to countenance such a view. He says:

> There are two different cases in the use of the word "I" (or "my") which I might call "the use as object" and "the use as subject". Examples of the first kind of use are these: "My arm is broken", "I have grown six inches", "I have a bump on my forehead", "The wind blows my hair about". Examples of the second kind are: "*I* see so-and-so", "*I* hear so-and-so", "*I* try to lift my arm", "*I* think it will rain", "*I* have a toothache". One can point to the difference between these two categories by saying: The cases of the first category involve the recognition of a particular person, and there is in these areas the possibility of an error, or as I should rather put it: The possibility of an error has been provided for. . . It is possible that, say in an accident, I should feel a pain in my arm, see a broken arm at my side, and think it is mine, when really it is my neighbour's. And I could, looking into a mirror, mistake a bump on his forehead for one on mine. On the other hand, there is no question of recognizing a person when I say I have a toothache. To ask "are you sure that it's *you* who have pains?" would be nonsensical.[10]

But does the fact that in regard to some 'I'-sentences the possibility of error has been provided for, imply that in those sentences the role of 'I' is to refer to an object? I do not see that it does. Nor does it follow that in an 'I'-sentence where no possibility of error has been provided for, 'I' means a thing called 'the subject'.

A point made several times by Anscombe is that if 'I' is a name, or a demonstrative, or some other kind of referring expression the role of which is to make a 'singular reference' to an object or thing, then there must be some 'conception' that connects 'I' with the object.

> The use of a name for an object is connected with a conception of that object. And so we are driven to look for something that, for each "I"-user, will be the conception related to the supposed name "I", as the conception of a city is to the names "London" and "Chicago", that of a river to "Thames" and "Nile", that of a man to "John" and "Pat". Such a conception is requisite if "I" is a name . . . .(pp. 51–52)

The requirement of a 'conception' is not removed if 'I' is assumed to be a demonstrative instead of a name.

> Assimilation to a demonstrative will not—as would at one time have been thought—do away with the demand for a conception of the object indicated. For, even though someone may say just "this" or "that", we need to know the answer to the question "this *what*?" if we are to understand him; and he needs to know the answer if he is to be meaning anything. (p. 53)

Anscombe adds the following remarks in a footnote:

> This point was not grasped in the days when people believed in pure ostensive definition without the ground's being prepared for it. Thus also in those days it was possible not to be so much impressed as we ought to be, by the fact that *we can find no well-accounted-for term corresponding to "I"* as "city" does to "London". It was possible to see that there was no 'sense' (in Frege's sense) for "I" as a proper name, but still to think that for each one of us "I" was the proper name of an 'object of acquaintance', a *this*. What *this* was could then be called "a self", and the word "self" would be felt to need no further justification. (p. 53; footnote; my emphasis)

It seems right to say that we are unable to specify any conception or well-understood general term that corresponds to 'I', such that in saying 'I. . . .' one means 'I' to refer to something falling under the conception or term. The conception 'human body' is well enough understood; but in a state of sensory deprivation, such as we previously imagined, one might think, 'Perhaps I am bodiless'; and this thought would be incoherent if the conception 'human body' corresponds to 'I'. The same objection would apply to the conceptions of 'human being' or 'person'; for in our usual employment and understanding of these expressions, to be a human being or person includes having the human physical form. Descartes's thought 'I have no body', actually entailed the thought 'I am not a human being'.

If we turn to 'self', 'soul', or 'mind', as conceptions

corresponding to 'I', the fact is that these expressions are not well understood. I should hardly know what I was saying if I either asserted or denied that I am a self, soul, or mind.

There are occasions on which I say to others, who want to know who I am: 'I am N.M.', or 'I am the youngest son of Charles M.'. These are true statements. But Anscombe makes the striking remark that although such statements are true they are not *identity* statements. She says:

> If I am right in my general thesis, there is an important consequence—namely, that "I am E.A." is after all not an identity proposition. (p. 60)

I think her point is that in a genuine identity proposition a distinctly conceived subject is connected with a distinctly conceived predicate. For example, 'Descartes was the author of *Meditationes de prima philosophia*'. Here the name 'Descartes' refers to a distinctly conceived subject—a human being, a Frenchman, who lived in the seventeenth century, did military service, invented analytical geometry, and so on. But when Descartes introduced himself to others by saying, 'I am René Descartes', he was not making an identity statement, although what he said was true.

If I were to lose my memory (perhaps from a concussion) so that I no longer knew who I was, it could be imagined that subsequently I became convinced by evidence and testimony presented by others, that my name is 'N.M.', that I was born in such-and-such a place of such-and-such parents, that the course of my life has been so-and-so. I should have found out who I am; I should have *learned my identity*. There might be a moment of sudden conviction at which I exclaimed: '*So*; *I* am N.M.; my parents were X and Y; the course of my life has been thus-and-so!'

I have found it tempting to think that this statement, made at the moment of my realization of my identity, would be a genuine identity proposition. But is this right? What reason is there for thinking that the use of 'I' in this statement is governed by a conception of an object, any more than it is in the utterance 'I am in pain', or in the thought that might occur in a condition of sensory deprivation, 'Perhaps I am bodiless'? I can see none.

Suppose that from extreme privation and suffering I no

longer knew who I was. Lying on the ground amidst other enfeebled people, and feeling a savage thirst, I cry out, 'Water! Water!'. An attendant, looking around, says, 'Who wants water?' I might call out 'I'; or might instead raise my hand. By uttering 'I', or by holding up my hand, I would equally be drawing attention to myself. But neither the hand gesture nor the utterance 'I', would imply a conception of myself that distinguished me from other persons. My utterance 'I' in response to the question, would show that I still *retained a mastery of the use of* 'I', even though I had no conception of who I was.

My conclusion is that Anscombe is entirely right in her contention that 'I' is not a referring expression, despite our inclination to think it is. This is so, not only in respect to the first person, present tense *Ausserungen* ('I am in pain', 'I am thirsty'), to which Wittgenstein has drawn attention; but also in respect to statements such as 'I am sitting in a chair', where there is the possibility of error; and also in statements by which I intend to inform others of my identity ('I am N.M.', 'I am the youngest son of Charles M.'). *Nowhere* does 'I' have the role of designating or meaning a distinctly conceived object *or* subject.

Notes

1 G. E. M. Anscombe, 'The first person', in *Mind and Language*, ed. S. Guttenplan (Clarendon Press, Oxford, 1975).
2 D. Hume, *Treatise*, Book I, Section VI.
3 T. Reid, *Essays on the Intellectual Powers*, Essay III, Ch. IV.
4 G. E. M. Anscombe, *op.cit.* p. 49. Hereafter I will put page references to Anscombe's essay into the text.
5 B. Russell, 'The philosophy of logical atomism', in *Logic and Knowledge*, ed. R. C. Marsh (Allen & Unwin, London, 1956), p. 243.
6 L. Wittgenstein, *Tractatus Logico-Philosophicus*, trans D. F. Pears and B. F. McGuinness (Routledge & Kegan Paul, London, 1962) 3.2–3. See L. Wittgenstein's *Philosophical Investigations*, trans G. E. M. Anscombe (Basil Blackwell, Oxford, 1973), para. 46.
7 L. Wittgenstein, *Philosophical Investigations*, para. 404.
8 L. Wittgenstein, *The Blue and Brown Books* (Basil Blackwell, Oxford, 1969), p. 68.
9 See L. Wittgenstein, *Philosophical Investigations*, *op. cit.* para. 405.
10 L. Wittgenstein, *The Blue and Brown Books*, pp. 66–67.

[3] INDIRECT REFLEXIVES AND INDIRECT SPEECH

J. E. J. Altham

In 'On beliefs about oneself',[1] Geach drew attention to predicates containing an indirect reflexive pronoun. An example he gave was '—believes that he himself is clever', 'he himself' here being the indirect reflexive. He characterized this pronoun as an *oratio obliqua* proxy for the first-person pronoun of *oratio recta*. He also said that if we say of a number of people that each of them believes that he himself is clever, we are not saying that they all believe the same proposition, as 'proposition' is commonly understood by philosophers. These remarks of his serve to introduce the main themes of this paper. In it I hope to make explicit some of the difficulties indirect reflexives present, both for accounts of indirect speech along Fregean lines, in which reference to propositions or other 'intermediate objects'[2] is essential, and for some accounts in terms of direct speech. The prospects for overcoming the difficulties look better, however, for the latter kind of account.

Use of 'he' as an indirect reflexive is a kind of use not to be assimilated to other uses of the pronoun. It is not, for example, a demonstrative, nor an ordinary bound variable. Nor does it stand proxy for a name or definite description. Its distinctness was fully brought out by Castaneda,[3] and has recently been reinforced by Anscombe.[4] Early in her essay, she is concerned to refute the idea that the first-person pronoun, 'I', can be explained as a referring expression by saying 'It's the word each one uses in speaking of himself'. If 'himself' is the ordinary reflexive, then one can speak of oneself without knowing that one is so doing, and in that use 'speaking of himself' does not explain 'I'. If, on the other hand, 'himself' is the indirect reflexive, then the explanation is back to front. For Anscombe holds that the indirect reflexive can be explained only in terms of the first person. The indirect reflexive is the reflexive of indirect speech, and the first person in terms of which it is to be explained is the first person of direct speech, so the claim is that with respect to this element, indirect has to be explained in terms of direct speech. It would be strange if this were so only for one element of indirect speech, so it is natural to generalize the claim to one that indirect speech in general is to be explained in terms of direct speech. The question then is how this is to be done.

First, however, here are a few reminders about the indirect reflexive. Its difference from the direct reflexive can be illustrated as follows. 'Reginald hurt himself falling downstairs' is logically equivalent to 'Reginald hurt Reginald falling downstairs'. Here 'himself' is a direct reflexive. But 'Reginald said that he himself had been to Venice' neither entails nor is entailed by 'Reginald said that Reginald had been to Venice'. Sometimes a 'he' in an indirect context is replaceable by its antecedent. An example would be 'Mary smiled at John, not knowing that the boy who had sneered at her was he'. Here we can say that what Mary did not know was that John was the boy who had sneered at her. Thus the 'he himself' in the previous example is not like this 'he'.

Suppose that, in 'Reginald said that he himself had been to Venice', we regard 'he himself' as a referring expression. Then it might be said that the possibility of changing the truth-value of the whole by replacing the pronoun with 'Reginald' is simply an instance of the well-known phenomenon, that terms coextensive in extensional contexts may not be intersubstitutable in non-extensional ones. But there must be more to it than that. Here is one further peculiarity. There is no way at all of replacing the indirect reflexive, and this has the consequence that we cannot specify what Reginald said by any complete sentence, so long as we remain in indirect speech. 'He himself had been to Venice' cannot itself be regarded as complete, for either the pronouns remain reflexive, in which case it is not complete, or the pronouns are otherwise used, in which case it no longer represents what Reginald said. Normally, of course, in an instance of '*a* said that *p*', '*p*' is a complete sentence, or one that can be completed from the context, as in the example of Mary and John in the previous paragraph. If we wish to find a complete sentence contained in 'Reginald said that he had been to Venice', we have to switch to direct speech. Then the relevant sentence is obvious, namely 'I have been to Venice'.

So much by way of introduction. The task now is to look at Frege's theory of indirect discourse in the light of indirect reflexives. According to Frege, terms have both sense and reference. The notions are introduced first for extensional contexts. For a proper name in such a context, the reference is what the name names, and the sense is the way in which the reference is given by that name. In indirect discourse, terms do not have their ordinary sense and reference. They rather refer to what would, in ordinary contexts, be their sense. Thus the indirect reference of a term is its ordinary sense. Frege also suggested that in an indirect context a term had indirect sense. This would be the way in which the indirect reference was presented. The notion of indirect sense will not, however, be of much concern in this essay.

In Frege's theory, a term has to have a sense in order to have a reference. Furthermore, unless a term has an ordinary sense, it cannot have an indirect reference, since indirect reference just is ordinary sense. In order to apply the theory to the analysis of a particular sentence in indirect discourse, it is necessary to identify logical units, the items to be considered as having sense and reference. So the first problem is to see how this is to be done. A simple case would be 'John said that Mary loves Peter'. Here 'Mary', '—loves . . .', and 'Peter' can be taken as units in the subordinate clause, and the analysis can proceed as follows. In this context, 'Mary' stands for the sense of 'Mary', i.e. the sense 'Mary' has in ordinary contexts, 'Peter' for the sense of 'Peter', and '—loves . . .' for the sense of '—loves . . .'. The whole subordinate clause then stands for the sense of 'Mary loves Peter'. (This is very sketchy, and ignores problems of interpretation. I hope these shortcomings do not matter.) Can this pattern of analysis be repeated for a simple sentence in indirect discourse containing an indirect reflexive? Consider this one: 'John said that he loved Mary'. Here the reference of 'Mary' can be taken to be the (ordinary) sense of 'Mary', but what about 'he', when this is taken as an indirect reflexive? It looks like a logical unit in the subordinate clause, but it is hard to see what, if so, its ordinary sense, and hence its indirect reference, might be.

The first problem here is that strictly, the word 'he', used as an indirect reflexive, *cannot* have an ordinary sense. For by its definition as indirect, it does not occur with this use in direct contexts. It might then be suggested that it has, as what would be its ordinary sense, the actual ordinary sense of the pronoun which corresponds to it in direct speech. This is the first-person pronoun 'I'. So the suggestion is that the reference of the indirect reflexive is the sense of 'I', the sense 'I' has in direct speech. Anscombe, however, has shown[5] that the attribution of a Fregean sense to 'I' leads to insoluble difficulties. A Fregean sense of 'I' would be a way in which its ordinary reference was given, and this is, in her expression, a conception under which one latches onto what is referred to. But I take her to have shown that 'I' contains no conception under which one latches onto anything. The suggestion that the reference of 'he' is the sense of 'I' therefore does not work, since there is no such thing as the sense of 'I'.

It is not easy to perceive what else the reference of the indirect reflexive might be—its *indirect* reference that is—for indirect reference is explained only through the notion of customary sense. The natural candidate for this customary sense fails, through there not being any such thing. The antecedent of the reflexive, 'John' in my example, does have a customary sense; but the indirect reference of 'he' cannot be this sense, since if it were, 'John' could be substituted without change of truth-value for 'he'. This substitution, however, has already been seen to allow the possibility of change of truth-value.

If 'he' (as an indirect reflexive) did have an indirect reference, it would have to be a sense, and it seems that it could not be any kind of sense other than that which contains a conception under which one can latch onto something. On the other hand, it could *not* be that kind of sense, since the correspondence between 'I' and 'he' is such that if 'I' has no (Fregean) sense, 'he' could not either. The point is that the truth of 'John said "I love Mary"' is sufficient for the truth of the report 'John said that he loved Mary'. (I ignore such qualifications as that John knew what he was saying, and background assumptions such as that 'I love Mary' has meaning. These do not seem relevant in the immediate context.) For it to be true that John said that he loved Mary, John does not have to have used a term that referred to himself under any conception. He could merely have used 'I'. So 'he' cannot get an indirect reference from any customary sense at all, since it contains no conception under which an object of reference might be latched onto.

From this it follows that 'he' cannot, consistently with Frege's theory, be a logical unit in 'John said that he loved Mary'. But there seems no other way of analysing the subordinate clause in Fregean theory. Another way of seeing the difficulty is to take up a point made earlier, that no complete sentence can be found contained in the subordinate clause. In normal cases the Fregean analysis issues in assigning (as indirect reference of the whole subordinate clause) the sense of words that form a complete sentence. But where an indirect reflexive is involved, there is no such sentence available.

It is worth comparing this case with one that is superficially similar, but which a Fregean can deal with. Consider the sentence 'Jane said that I bungled'. Now Jane certainly did not say 'I bungled'. It is more likely that she said 'Jimmy bungled', or 'Dr Altham bungled'. The sentence says that Jane referred to me, but it does not enable us to say what words she used to do so, nor through what conception she latched onto me. This, however, need not worry a proponent of Frege's theory. For while it is not said what conception it was through which she referred to me, we know that there must have been one. (Even a demonstrative reference involves some such conception.) So the sentence can be analysed as something like 'There is a sense *s* that determines me as reference, and Jane said that *s* bungled'. A similar analysis does not work for the indirect reflexive, since there need be no such sense through which a man refers to himself.

An attempt to treat 'he' as a bound variable would come up against a similar objection. Thus, suppose one tried to start analysing 'John said that he loved Mary' as 'There is someone who is John and who is said by John to love Mary'. This does not work, for quantification into an indirect context—in Fregean theory—involves quantifying over senses, and once again no appropriate senses are available.

It would be most unsatisfactory and implausible to treat cases of indirect speech not involving indirect reflexives in Frege's way, merely regarding cases that do involve these reflexives as a separate category that need not be integrated into the Fregean account. So, unless the latter cases can be incorporated, Fregean theory must be considered as having received a serious blow.

Kaplan[6] advises that we should distinguish between the principle that expressions in indirect contexts do not have their ordinary reference (which he holds to be true), and the principle that expressions in indirect contexts refer to their ordinary senses (which he holds to be in general false). He thinks that if a particular suggestion for the 'intermediate objects' is found wanting, we should go on looking for others. However, the possible variety of intermediate objects does not appear to be very wide. What else, one might wonder, could be involved other than expressions, what these stand for, and what they mean? Kaplan himself favours an analysis in which the subordinate clause in indirect speech gives place to a quoted sentence. This latter has to be in direct speech, and this, despite what Kaplan says, seems to take us far from the spirit of Frege's own theory. It will be more appropriate to say something about quotation later on in this essay.

So far the discussion seems to have lent some support to Geach's remarks with which I started. The ordinary philosophical sense of 'proposition', as the sense of a declarative sentence determined by the senses of its parts, does not seem to be a useful instrument in the present connexion. Moreover, the difficulties seen so far may reasonably encourage a closer look at the relation with direct speech. Here it may be helpful to mention some puzzling features of Davidson's account,[7] when applied to indirect reflexives.

Davidson's theory regards a sentence of indirect speech as in effect two sentences. I take his example 'Galileo said that the earth moves'. Davidson's analysis regards an utterance of this as two utterances. One of these is 'Galileo said that'. Here 'said' is a two-place predicate, and 'that' is a demonstrative. The demonstrative refers to the second utterance, which is 'The earth moves'. One crucial difference from accounts in the Fregean tradition, and some others, is that 'from a semantic point of view the content-sentence in indirect discourse is not contained in the sentence whose truth counts'.[8] The one whose truth counts is of course the former, 'Galileo said that', and 'The earth moves' is semantically no part of it.

Introducing his analysis, Davidson reaches 'Galileo said that the earth moves' as essentially an abbreviation of the following pair:

The earth moves.

($\exists x$) (Galileo's utterance x and my last utterance make us samesayers).

Two people samesay when the utterance of one says the same as an utterance of the other. Davidson does not explain the conditions for that. The fuller version of the analysis makes the truth-conditions clear, provided that we waive the problem of truth-conditions for saying the same thing. (This is a large waiver, rather like a poor man's waiving a right to the £100,000 Premium Bond prize.)

Now let us apply this pattern of analysis to a sentence involving an indirect reflexive, for example 'John said that he loved Mary'. The obvious thing to do is to recover the sentence in direct speech corresponding to the subordinate clause, and make that the first utterance in the analysis. Then the second utterance will be like the second one displayed above, except that 'John' replaces 'Galileo'. So the result is:

I love Mary.

($\exists x$) (John's utterance x and my last utterance make us samesayers).

However, as Davidson explains his analysis, this attempt is clearly wrong. For he thinks that in making the utterance that corresponds to the subordinate clause 'I speak for myself, and my words refer in their usual way'.[9] Hence if I make the utterances of the analysis, I speak of myself in uttering 'I love Mary'. But of course the utterance being analysed 'John said that he loved Mary', in no way speaks of me, but only of John and Mary. So this attempt is incorrect.

A second try would be:

He loves Mary.

($\exists x$) (John's utterance x and my last utterance make us samesayers).

Here the question is how 'he' is supposed to occur. As the utterance 'He loves Mary' is, in the analysis, semantically separate from the one that follows it, it could be made alone. The 'he' in it cannot then still be an indirect reflexive, since 'He loves Mary', so understood, can only occur as a fragment of some more inclusive utterance. The 'he' must then be either a proxy for 'John', or a demonstrative, but in either of these interpretations, the utterance 'He loves Mary' does not samesay with John's utterance. For, first, John said 'I love Mary', and this is significantly different, even in John's mouth, from 'John loves Mary'. Secondly, since 'I' does not serve to make a demonstrative reference, or at least not in the way that 'he' does, the utterance 'He loves Mary' in its other interpretation is still significantly different from John's own. Here again I am appealing to the point that a demonstrative reference involves a conception of what it refers to, and this is not true of 'I'.

The basic problem here can be put in a nutshell like this: when John utters a sentence containing 'I', thereby speaking of himself, he speaks of himself in a way in which it is impossible for anyone else, speaking for himself as Davidson stipulates, to speak of him. So nobody else can samesay with John's utterance of 'I love Mary'. It would be useless to counter this with an assertion that the criteria for samesaying being used are unrealistically strict. For one thing, the reasons for not insisting on ubiquitous sharpness of answers to questions about synonymy have at bottom to do with the thesis of the indeterminacy of translation, which is not in question here. The present problem is mercifully a more local one. For another, the objections to the proposed analyses rest on simple recognition that they are clearly wrong, that for example, if John said that he loved Mary, it would often be incorrect to report him as having said that John loved Mary.

It is now time to link the present problem, the impossibility of my utterance sufficiently matching John's in the right way to make us samesayers, with another puzzling feature of Davidson's account. Suppose we ask what sort of speech-act I make when I utter 'The earth moves', as one of the pair of utterances which analyse 'Galileo said that the earth moves'. What illocutionary force would it normally have in this context? I shall assume that the context includes my *asserting* the whole sentence 'Galileo said that the earth moves'. This assumption is merely for definiteness and simplicity. Davidson does not himself require any particular force for the utterance referred to. The utterance introduced by 'that', he says, may be 'done in the mode of assertion or of play'.[10] This phrase conceals an important point, which may be brought out by comparing

Galileo said that the earth moves.

and

Galileo was aware that the earth moves.

For the sake of illustration, I assume that 'aware that' is to be subjected to a pattern of analysis similar to that for 'said that'. Thus we should have:

The earth moves.
Galileo was aware of that.

The point here is that even if the former of this pair is not itself uttered assertively, the assertion of the latter will (in this case retrospectively) give that utterance assertoric force. So the difference in assertoric content between '*a* said that *p*' and '*a* was aware that *p*' can be brought out, in Davidson's analysis, not through a difference in the force of the utterance '*p*', but through the difference between saying and being aware. This perhaps lends support to the idea that the illocutionary mode of utterance of the content-sentence does not really matter. Perhaps, on the other hand, it does matter. For suppose I assert both 'The earth moves' and 'Galileo said that'. This double assertion cannot be offered as an analysis of the assertion 'Galileo said that the earth moves'. It corresponds rather to 'Galileo said that the earth moves, and the earth does move'. Moreover, it is impossible so to utter 'Galileo said that the earth moves' as to assert that the earth moves. So unless something is said about the way in which the content-sentence is supposed to be uttered, the analysis can be used for a greater variety of speech-acts than the sentence to be analysed.

What is needed is for the utterance of 'the earth moves' to occur unasserted. But merely to require that is to leave too much obscure. Logically, an occurrence is unasserted if it is in effect a component of a longer sentence. On Davidson's analysis the content-sentence is not a component. A different notion of unasserted occurrence is needed, the natural model for which is play. So the thing to try out is the idea that in indirect discourse the content-sentence occurs in a way analogous to some kind of playful occurrence.

There is one mode of play that promises to resolve our

puzzles about reflexives and the first person. The main problem was that if John uses 'I', I cannot quite say what he says, unless I am John. This problem seemed insoluble for Davidson's analysis, but the insolubility depended on the assumption that in uttering the content-sentence I speak for myself. So suppose this assumption is dropped, and it is supposed instead that in reporting what John said, I am speaking for *him* rather than for myself. The idea is this. When I say 'I love Mary, John said that', my utterance of 'I love Mary' is like an impersonation of John. I am in a sense and to a certain degree putting on an act of being him. The second utterance 'John said that' specifies to the audience who it is that I am impersonating.

This suggestion may seem bizarre, and perhaps 'impersonation' is a rather startling word for what I am getting at. To clarify matters, I want to consider some cases of reports of speech in *oratio recta*. First, consider one of those friends of Socrates who is represented by Plato as having an amazing memory for speeches and conversations. Let us call him Hetairos. Hetairos has just heard Gorgias make a speech when Glaucon comes up, and is very disappointed to have missed it. Hetairos tells him not to mind, because he Hetairos can remember the speech word for word and will repeat it to Glaucon, which he thereupon does, starting by saying, 'His speech went like this', and then launching into it. Now if Hetairos were to write it down instead, and were following modern conventions, after 'this' he would put inverted commas, with another pair after the last word of the speech. A modern logician, looking at this, and adopting *his* understanding of quotation marks, would regard the whole string inclusive of the inverted commas as a name of what Gorgias said. But that is surely not how Hetairos meant them. What he wrote represents on paper what he uttered, and surely he did not utter a name of Gorgias's speech. Rather, he repeated Gorgias's speech, that is, he produced another token of it, not a token of a name of it.

One possibility is that Hetairos is an excellent actor, and says to Glaucon, 'Never mind, you just sit there, watch and listen, and I'll do Gorgias making his speech for you'. He then imitates Gorgias's posture, gestures, tone of voice and so forth. He tries to be as much like Gorgias as possible. This would be an attempt at full impersonation. A less able Hetairos might be content only to repeat Gorgias's words. This might be called impersonation only with respect to Gorgias's words. Finally, a Greek with a slightly less amazing memory might be able to reproduce only the content of Gorgias's speech. This could be called impersonation only with respect to what Gorgias said. He need not, of course, in this third case, resort to the grammatical constructions of indirect speech, for one who uses direct speech to report another does *not* need to claim that he is using exactly the same words. But in this third case, Hetairos is doing something quite interchangeable with the use of *oratio obliqua*.

If we start with the idea of full impersonation, and then little by little drop from it imitation in one respect after another, while keeping to imitation with respect to content of what is said, we do not reach the idea of something different in kind from imitation. Certainly we do not switch at any stage from reproducing what is said to producing a name of it. What we do reach is something having just the standard function of indirect speech.

Instead of 'impersonation', I could use an expression of Geach's, quoted in the first paragraph. Hetairos *goes proxy* for Gorgias in repeating his speech. However, this could mislead. If a man votes by proxy, the vote is usually valid. It has the same consequences as if he had voted in person. An utterance by proxy, in the sense I intend, does not. A report in *oratio recta* of a command is not, for example, a command. So it is preferable to say that it impersonates one. Commands are indeed instructive in this connexion. 'Commanding that' and 'commanding to' must presumably be analysed in the same style as 'saying that'. Applying this style to 'John commanded that Mary should go', it is first necessary to recover the *oratio recta* sentence. This is 'Go, Mary'. The second utterance is just 'John commanded that'. Now, notoriously, commands do not appear as logically unasserted. 'Go, Mary' can equally not be asserted as it occurs in an analysis of 'John commanded that Mary should go'. Once these possibilities are excluded, it seems natural to regard the utterance of 'Go, Mary' as a representation of John's performance in issuing the command.

At this point it seems appropriate to return to the issue of quotation. I have already remarked that there is one use of reports in *oratio recta* in which no claim is made to give the exact words of the speaker reported. Clearly an *oratio recta* report, in English, of a speech made only in French by M Giscard d'Estaing is both possible—it happens—and does not give the French President's actual words. His actual words are neither used nor mentioned. Is there then any reason left to suppose that the use of quotation marks in such *oratio recta* reports signals mention of the words within them? For why should we mention any words that are not the President's? There is no point in doing so. We are using *oratio recta* to report what he said, not the words he used to say it.

Logicians and philosophers have great need of a convention for distinguishing between use of a word to talk about a thing and use to talk about itself, and have established the use of quotation marks for that purpose. The great importance of avoiding confusion has led to some neglect of other uses of quotation marks, and other senses of 'quote'. To take a simple case, if an old-fashioned schoolmaster says that *mens sana in corpore sano* is what he tries to inculcate in his pupils, he is quoting Juvenal, but is not mentioning his words, but using them to talk about a healthy mind in a healthy body. To quote is frequently, as the dictionary says,[11] to repeat a passage of someone else's. The quotation marks, if used, serve as an indication that the passage is borrowed, not to signal mention.

I hold that the use of quotation marks in *oratio recta* reports, where it is understood not to be required that the report gives the speaker's actual words, does not signal mention of words. The quotation marks are the device used in writing to signal a change in speech-act, from what would normally be assertion to what I am calling impersonation with respect to the content of what was said.

I can now summarize and draw the threads together. To explain indirect speech in terms of intentional entities such as propositions looked unpromising, because there is no intentional content in the indirect reflexive. There is no intentional content in that because there is none in the first-person pronoun 'I'. It seemed better therefore to explore indirect reflexives, and through them indirect speech, by exploring an analysis in terms of direct speech, here following

Anscombe's advice. Davidson makes the subordinate clause in indirect speech a semantically separate utterance. This it can only be if it is put back into direct speech. But then there is no way of putting the indirect reflexive back into direct speech while retaining Davidson's assumption that I am speaking for myself. If that assumption is dropped, *then I can use 'I' to talk about another*, by impersonating him. This is the core of the solution. For Davidson's analysis to work, moreover, it must be required that the utterance corresponding to the subordinate clause should occur unasserted, and the idea of an impersonating occurrence fits this requirement. Finally, *oratio recta*, on one understanding, has the same function as *oratio obliqua* in reporting speech, and so must be amenable to the same analysis. This is seen to be possible once it is seen that quotation marks do not always work the same way, and that in this use of *oratio recta* there is no reason to suppose that they signal the mention of expressions. The way then seems open to supposing that they signal a change of speech-act, to impersonation. A paratactic and a quotational account can thus be made to coincide.

Notes

1 P. T. Geach, 'On beliefs about oneself', *Analysis*, 1957–8, reprinted in P. T. Geach, *Logic Matters* (Basil Blackwell, Oxford, 1972), pp. 128–9.
2 D. Kaplan, 'Quantifying in', in *Words and Objections* eds. D. Davidson and J. Hintikka (Reidel, Dordrecht, 1969), p. 213.
3 H. Castaneda, ' 'He': A study in the logic of self-consciousness', *Ratio*, 8 (1966), 130–157.
4 G. E. M. Anscombe, 'The first person', in *Mind and language*, ed. S. Guttenplan (Clarendon Press, Oxford, 1975), pp. 45–66.
5 G. E. M. Anscombe, *op. cit.*
6 D. Kaplan, *loc. cit.*
7 D. Davidson, 'On saying that', in *Words and Objections*, eds. D. Davidson and J. Hintikka (Reidel, Dordrecht, 1969), pp. 158–174.
8 D. Davidson, *op. cit.* pp. 170–171.
9 D. Davidson, *op. cit.* p. 169.
10 D. Davidson, *op. cit.* p. 171.
11 See, for example, *The Concise Oxford Dictionary*, sixth edition, ed. J. B. Sykes (OUP, Oxford, 1976).

[4] THE INDIRECT REFLEXIVE
Roderick M. Chisholm

1. Introduction

I shall discuss the 'problem of the indirect reflexive' and its relation to what is expressed by means of the first-person pronoun. My topic, then, is essentially the same as that of Professor Anscombe, in her paper 'The first person.'[1] But where she suggests that the indirect reflexive 'can be explained only in terms of the first person', I shall suggest that the use of the first person can be explained only in terms of the indirect reflexive.

2. The Problem of Indirect Reflexives

Let us begin by citing two examples from Anscombe. She notes, first, that it is one thing for Descartes to doubt the identity of *Descartes* with Descartes and quite another thing for him to doubt the identity of *himself* with Descartes.

I will quote Anscombe's second example:

> "When John Smith spoke of James Robinson he was speaking of his brother, but he did not know this." That's a possible situation. So similarly is "When John Smith spoke of John Horatio Auberon Smith (named in a will perhaps) he was speaking of himself, but he did not know this." If so, then 'speaking of' or 'referring to' oneself is compatible with not knowing that the object one speaks of is oneself.

Anscombe goes on to say that the expression 'He doesn't realize the identity with himself' is not the 'ordinary' reflexive but 'a special one which can be explained only in terms of the first person'. To understand this reflexive, let us consider its use in clauses expressing the objects of believing. We may contrast the three locutions:

P—The tallest man believes that the tallest man is wise

Q—There is an *x* such that *x* is identical with the tallest man and *x* is believed by *x* to be wise

S—The tallest man believes that he himself is wise

The distinction between *P* and *Q* is familiar; and it is generally agreed that *P* does not imply *Q*, and *Q* does not imply *P*. But it is also clear that *S* is logically independent of *P* and that although *S* implies *Q*, *Q* does not imply *S*.[2] Under what conditions might it be the case that *Q* is true and *S* is false?

If *S* is false, then the tallest man cannot sincerely say 'I believe that I am wise'. Suppose now that the tallest man reads the lines on his hand and takes them to be a sign of wisdom; he doesn't realize the hand is his; and he is unduly modest and entirely without conceit. And so, although he cannot sincerely say, 'I believe that I am wise', he does correctly express his conclusion by saying, 'Well, *that* person, at least, is wise'. One might thus be led to ask: can we paraphrase the 'he, himself' locution of *S* in such a way that the result will not resist the ordinary notation of quantification?

It will be useful to have neutral terms for the two types of reflexive relation. In place of the grammatical term 'indirect reflexive', let us use the term 'emphatic reflexive' for the 'he himself' reflexive such as *S*; and let us use 'non-emphatic reflexive' for those reflexive relations such as *Q* that do not imply the 'he himself', or emphatic, reflexive. Thus 'There exists an *x* such that *x* believes himself to be wise' will express an emphatic reflexive, and 'There exists an *x* such that *x* believes *x* to be wise' will express a non-emphatic reflexive.

Perhaps it will be agreed that the distinction between the two types of reflexive is one that holds only in intentional or psychological contexts. Now there are two ways of interpreting the significance of this distinction. In either case, we ask 'Why is it that the non-emphatic reflexive (for example, *Q*) does not imply the corresponding emphatic reflexive (for example, *S*)?' But in the one case, we will trace the failure of implication to certain peculiarities of the emphatic reflexive. And in the other case, we will trace it to certain peculiarities of the non-emphatic reflexive. If we take the second approach, we will deny that the emphatic reflexive presents us with any unique logical structure. We will say that the failure of implication is due, rather, to certain familiar facts about intentionality—as exhibited in the *non-emphatic* reflexive.

I suggest that, by exploring the possibility that the second approach is correct, we may arrive at a view enabling us to understand the logic of the two types of reflexive. It will be instructive to remind ourselves of this fact: there are philosophers who (justifiably or unjustifiably) are not convinced of the validity of the distinction between the two types of reflexive; and what they are sceptical about is, not the existence of the emphatic reflexive, but the existence of the non-emphatic reflexive—they doubt that there is a sense of 'There exists an *x* such that *x* believes *x* to be wise' that does *not* imply 'There exists an *x* such that *x* believes itself to be wise'. What is peculiar in the distinction, they might say, is the assumption that there is a reflexive which is *not* emphatic.

Consider, then, the possibility that, in the case of non-psychological reflexives, *all* reflexives are emphatic. In the case of motors, say, it will not matter whether we say 'There is an *x* such that *x* refuels *x*' or 'There is an *x* such that *x* refuels itself'; there is no non-emphatic reflexive here.

3. Some Attempted Solutions

It is obvious that the emphatic reflexive is closely related to the use of first-person *sentences*. And it is commonly assumed that first-person sentences express first-person *propositions*. Our treatment of the emphatic reflexive, therefore, will presuppose a theory about first-person propositions.

There is clearly a difference in meaning between the sentences, 'I am wise' and 'Someone is wise'. The former would seem to imply the latter but not conversely. What is the difference? We may say at least this much: 'I am wise' expresses something that 'Someone is wise' does not express. It is tempting to say: 'The *proposition* expressed by the former entails that expressed by the latter and not conversely'. One could then add: 'When I state that I am in pain, I am expressing the proposition that I am in pain, just as you, when you state that you are in pain, are expressing the proposition that *you* are in pain'. This presupposes that there *is* a certain proposition that I express with the words 'I am in pain' and presumably *another* proposition that you express with the same words. Obviously there are 'I'-sentences, first-person sentences. But are there thus 'I'-propositions, first-person propositions?

Of the theories of the emphatic reflexive that have been

formulated so far, each has its characteristic interpretation of first-person propositions. Each theorist suggests that he can express his own first-person propositions by using the first-person pronoun. And each theorist seems to despair of being able to express any first-person proposition of any person other than himself. We may distinguish four theories:

(i) There is Anscombe's theory which she summarizes as follows: '"I am this thing here" is, then, a real proposition, but not a proposition of identity. It means: this thing here is the thing, the person (in the 'offences against the person' sense) of whose action *this* idea of action is an idea, of whose movements *these* ideas of movement are ideas, of whose posture *this* idea of posture is the idea.'[3] She thus attempts to explicate *her* use of the first-person pronoun in terms of 'this'. It is clear that she cannot explicate *my* use of 'I' this way. And I think she might concede that she cannot grasp my 'I'-propositions at all.

(ii) Castaneda also assumes that there are first-person propositions. He tells us that, when a person uses an 'I'-sentence, then he is expressing a first-person proposition which 'is different from every third-person proposition about him, and, of course, different from any third-person proposition about anything else'.[4] Castaneda thus seems to suggest the view that he could never express *my* 'I'-propositions, and I believe he would say that, strictly speaking, he cannot even grasp them.

(iii) I have defended the view that, for each person, that person's use of 'I' is such that he is its referent and his individual essence or haecceity is its sense.[5] I suggested, too, that no one is able to grasp the 'I'-propositions of any other person.

(iv) I have also proposed an alternative theory, making use of the concept of *de re* belief and the concept of empirical certainty.[6] I had suggested that one might define 'x believes himself to be F' by saying 'It is certain for x that there is a y such that y believes y to be x'. According to this interpretation, the person who believes himself to be wise is one for whom it is certain that there is something which is believed to be wise. For I cannot be certain that there is someone who is *believed to be* wise without being that

> person. But this approach has the following difficulty. What rule of evidence could we give which would state the conditions under which it would be certain for a given subject that there is someone who is believed to be *F*? The property of being *F* would seem not to be self-presenting, for one can have it (if one is believed by someone else to be *F*) without being certain that it is had. And, more important, it is possible for a person to be believed by himself to be *F* without it being the case that it is certain for him that there is someone who is believed to be *F*. Shall we say 'The proposition that someone is believed to be *F* is certain for *S*, provided only *S* believes himself to be *F*'? Then we will be left with the indirect reflexive.

What is it, then, that we grasp when, as is assumed, we grasp our own 'I'-propositions?

4. Are there First-Person Propositions?

Let us recall the nature of propositions (*Gedanken*, or *Sachverhalten*). Whether or not anyone is in pain, there is the proposition, *someone is in pain*, just as there are the propositions, *there are horses* and *there are round squares*. If we take the term *proposition* in its most common philosophical sense, we will say that propositions are eternal objects having this nature: they are capable of being asserted or accepted or entertained.

Our question is: Is there, in this sense of 'proposition', a proposition corresponding to the expression 'I am in pain'—a different proposition for each person who could utter the words 'I am in pain'? If there are such first-person propositions, then, one might think, they could all be put in the form, 'The *F* is in pain', but with a different subject-term for each person. Let us consider the consequences of this assumption.

We can say, of propositions, not only that they imply other propositions but also that some of them imply certain *properties*. The concept of a proposition implying a property is simple enough: a proposition implies a certain property if the proposition is necessarily such that, if it is true, then something has the property. Thus the proposition, *the man in the corner is in pain,* implies such properties as being a man, being in a corner, and being in pain. (A proposition may also be said to *exclude* certain properties. Thus the proposition that there are no unicorns excludes the property of being a unicorn. This means that the proposition is necessarily such that, if it is true, then nothing has the property. I believe we may say that, for every contingent proposition *p* and every contingent proposition *q*, if *p* logically implies *q* and *q* does not logically imply *p,* then either: (a) *p* implies some property that *q* does not; or (b) *p* excludes some property that *q* does not.)

Now if there are first-person propositions, different ones for different people, then my 'I'-propositions would imply an identifying property that only I have and yours would imply an identifying property that only you have. Some philosophers—for example, Frege and Husserl—have suggested that each of us has his own idea of himself, his own *Ich-Vorstellung* or individual concept. And some of the things that such philosophers have said suggest the following view: the word 'I', in the vocabulary of each person who uses it, has for its reference that person himself and has for its sense that person's *Ich-Vorstellung* or individual concept. The difference between my 'I'-propositions and yours would lie in the fact that mine imply my *Ich-Vorstellung* and not yours, and that yours imply your *Ich-Vorstellung* and not mine.[7] Surely a simpler theory would be preferable.

I would suggest, then, that we consider this possibility: that the function of 'I'-sentences is *not* that of expressing 'I'-propositions. I believe that, by so doing, we will find a theory that accounts for the data we have been considering and that is considerably simpler, metaphysically, than its alternatives. For what may be said of 'I'-sentences and their relations to propositions may also be said of 'this'-sentences, 'there'-sentences, 'here'-sentences, and the like—where such words as 'this', 'there', 'here' are used as indicator words.

'But', one may object, 'sentences containing indicator words may express what is true or false. And they may stand in logical relations to sentences which, presumably, *do* express propositions. "I am sad" and "This is red", respectively, imply "Something is red" and "Something is sad". How can this be, if the latter sentences express propositions and the former sentences do not?' Clearly, our account of sentences and propositions must be one enabling us to answer this question.

5. An Interpretation of Emphatic Reflexives

It can be shown that the various senses of belief—*de dicto, de re*, and the emphatic reflexive (the 'believes himself to be *F*' locution)—can be reduced to a single belief concept. This single belief concept does not require an alteration of the standard theory of quantification, nor does it commit us to the being of any objects other than states of affairs, properties (and relations), and individuals.

Believing must be construed as a relation between a believer and *some* other thing; this much is essential to *any* theory of belief. *What* other thing, then? There are various possibilities: sentences, propositions, properties, individual things. The simplest conception, I suggest, is one that construes believing as a relation between a believer and a property—a property which he may be said to attribute to himself. Then the various senses of believing may be understood by reference to this simple conception.

Analogous observations may be made with respect to other attitudes—for example, desiring, knowing, hoping, intending. In each case, there is an elementary conception of the attitude in question in terms of which more familiar conceptions can be explicated. And this elementary type of attitude, as we shall see, will enable us to understand the relation between the indirect reflexive and such indicator words as 'I', 'here', and 'this'.

There is a basic sense of believing, then, and two other senses definable in terms of the basic sense. To distinguish the three senses, we could simply use prefixes, as in '*a*-believes', '*b*-believes', and '*c*-believes'. But I shall use the following three expressions:

(i) 'the property of being *F* is such that *x* directly attributes it to *x*';

(ii) 'the proposition that *p* is accepted by *x*'; and

(iii) 'the property of being *F* is such that *x* attributes it to *y*'.

The second and third locutions may be defined in terms of the first.

Our basic doxastic locution, for which we could also use the expression '*a*-believes', may be spelled out as:

> The property of being *F* is such that *x* directly attributes it to *y*.

The letter '*F*' is schematic and may be replaced by any predicative expression: for simplicity we omit reference to a particular time. The undefined formula contains the two variables, '*x*' and '*y*', but we shall make the following stipulation concerning the nature of direct attribution:

> For every *x* and every *y*, if there is a property such that *x* directly attributes it to *y*, then *x* is identical with *y*.

Our undefined locution, therefore, is to be distinguished from the more general *de re* locution which allows us to say 'There exists an *x* and a *y* such that *x* is other than *y* and *x* believes *y* to be *F*'. This latter, for which we shall use locution (iii) above, will be characterized below in terms of our undefined concept of direct attribution.

One may ask: 'What is the *object* of direct attribution?' If something *x* is such that it believes *x* to have the property of being wise, then *x* may be said to be *an* object of *x*'s belief, and the property of being wise may *also* be said to be *an* object of *x*'s belief. Using a traditional terminology we could say that in this example *x* is the *object* of his belief and the property of being wise is the *content* of his belief. And we shall in fact use this terminology: in the case of direct attribution as well as attribution generally, we shall say that the property attributed is the *content* of the attribution and that the thing *to* which the property is attributed is the *object* of the attribution. But there is no reason to suppose that there is still *another* thing, somehow involving both the individual thing and the property of being wise, which is properly called '*the* object' of *x*'s direct attribution. This, despite the fact that in such a case one can ask, 'And *what* is it that he believes?'

The 'he himself' locution of the emphatic reflexive may be introduced as an abbreviation of our undefined locution:

> D1—*x* believes that he himself is *F* = *Df*. The property of being *F* is such that *x* directly attributes it to *x*.

One may ask: 'Why not *begin*, then, with the "he himself" locution? Then you would not need to take the trouble to spell out a definition of it.' The answer is that our undefined locution does not involve any extension of the notation of quantification. We take it as undefined, therefore, in order to make it clear that there is nothing in our account requiring us to go beyond this notation.

'Does application of your concept of direct attribution imply that the believer stands in a special relation of acquaintance with himself?' It implies, as we have said, that there is one sense in which the believer can be said to be *the* object of his believing. And it also implies that nothing other than the believer can be the object of direct attribution.

'How, then, does one succeed in making himself the object of his direct attribution?' It is important to note that this strange question has its analogue in application to *any* adequate conception of believing. For any such conception will imply that *something* or other is the object of any believing. And so one can always ask, with respect to such an object: 'And how, then, does one succeed in making that thing the object of his believing?' The answer can only be: 'One just does.'

We now introduce the *de dicto* belief locution, for which we will use the expression 'accepts':

> D2—The proposition that p is accepted by $x = Df.$ The property of being such that p is such that x a-believes x to have it.

In the case of b-believing it *is* appropriate to ask 'What is *the* object of his c-believing?' For in this case his believing may take a proposition as its object.

To accept the proposition that all men are mortal, then, is to attribute directly to oneself the property of being such that all men are mortal. Does this mean, therefore, that the belief is 'really a belief about oneself'? Not in the ordinary sense of the expression 'to have a belief about oneself'. For, ordinarily, we would not use this expression unless one attributed a more restricted type of property to oneself—for example, the property of being wise. But the property of being such that all men are mortal is an all-or-nothing property: it is necessarily such that either everything has it or nothing has it. We could say that, in the ordinary sense of the expression 'to have a belief about oneself', one does not have a belief about oneself unless the property that one attributes to oneself is one which is *not* an all-or-nothing property.

Let us now turn to *de re* belief. Consider the locution

enabling one to express the fact that I have a belief with respect to you, or, in other words, that there is a certain property I attribute to you. If I can thus have a belief with respect to you, then there is a relation I bear only to you.

For *de re* belief, then, let us use 'attributes':

> D3—The property of being F is such that x attributes it to $y = Df$. Either (1) y is identical with x, and the property of being F is such that x believes himself to have it; or (2) there is a property H and an R such that: (a) x believes himself to have H; (b) x bears R only to y; and (c) H is necessarily such that: (i) whoever conceives it conceives the property of being F; and (ii) for every x, x has H if and only if x bears R to exactly one thing and that thing has the property of being F.

Suppose, for example, that Jones believes, with respect to Jimmy Carter, that he is a Democrat. In this case 'being F' would be replaced by 'being a Democrat'; 'H' might be replaced by 'the property of living in a country where the President is a Democrat'; and 'R' might be replaced by some such expression as 'living in a country in which the President is identical with'. (The example should not be taken to presuppose the average citizen is in fact in a position to have a belief about the President.)

The definiens above could also be put in Professor Anscombe's 'under a description' terminology. We could say: 'Under one of its descriptions y is believed by x to be F—namely, as being the thing that x bears R to.' ('Under one of his descriptions, Jimmy Carter is believed by Jones to be a Democrat—namely, that of being the President of the country in which he lives.') But such 'under a description' talk would simply be short for what we have spelled out above.

Our definition above expresses a somewhat latitudinarian conception of *de re* belief. The definition could be tightened up by specifying that x *knows* himself to bear the relation R to exactly one thing. ('Jones knows that he lives in a country where there is a single president.') This would require that we introduce a concept of '*a*-knowing' which is the analogue of '*a*-believing'; that is to say, we should introduce a definition of the type of knowledge that is correlative to direct attribution and thus might be thought of as 'justified true direct attribution'.

And this presents no difficulty in principle. Thus we might have:

D4—x knows himself to bear the relation R to exactly one thing = *Df.* (a) x bears the relation R to exactly one thing; (b) the property of bearing R to exactly one thing is such that x believes himself to have it; and (c) there is a p such that: (i) x knows p to be true; and (ii) believing himself to bear R to exactly one thing is at least as reasonable for x as believing p.

We may note that, if a person x thus has a belief about a thing y, it is possible for x to be several steps removed from y. For example, if a certain house happens to be the thing that I am looking at, if Jack is the thing that built the house, if Mary is the lady who is married to Jack, and if I know these things, then I may have a belief with respect to Mary. For in this situation I will know that there is just one person who is married to the person who built the house that I am looking at.

Let us now return to the two sentences, S—'The tallest man believes that he himself is wise' and Q—'There is an x such that x is identical with the tallest man and x is believed by x to be wise'. What these sentences come to is the following:

S'—There is an x such that x is identical with the tallest man, and the property of being wise is such that x directly attributes it to x.

Q'—There is an x such that x is identical with the tallest man, and the property of being wise is one such that x attributes it to y.

It is clear from our definitions that S' implies Q'; and we have no reason to believe that Q' implies S'. And indeed we may be tempted to affirm in this context, a more general thesis about intentionality. In application to believing, it is the thesis that one may attribute a property to oneself without *directly* attributing that property to oneself. Hence we have made out the difference between what is asserted by the two sentences S and Q without modifying the theory of quantification or extending our ontology beyond individuals, properties and relations, and propositions or states of affairs. We may now return to Anscombe's two examples.

It was one thing, she noted, for Descartes to doubt the identity of *Descartes* with Descartes, and another thing for Descartes to doubt the identity of *himself* with Descartes.

Possibly the first doubt was a doubt concerning an attribution that is not direct. Or possibly the first doubt concerned a proposition—the proposition, say, that the philosopher called 'Descartes' is identical with the philosopher called 'Descartes'. But the second doubt, in any case, would have been a doubt pertaining to an attribution that is direct. As for John Horatio Auberon Smith, when he spoke of himself without knowing that he was speaking of himself, he was expressing an attribution that was not direct and he had not made the corresponding direct attribution.

If this type of analysis proves to be correct, then it may readily be extended to more complex cases of the emphatic reflexive. Consider, for example, '*S* believes that *T* believes that he, *S*, is wise'. We may explicate this in the following way: 'The property of being such that *T* attributes to it the property of being *F* is one that *S* directly attributes to *S*.' (Here the locution '*x* *c*-believes *y* to be *F*' would be more convenient for our *de re* locution. For we could say, more simply, 'The property of being *c*-believed by *T* to be *F* is one that *S* directly attributes to *S*'.)

6. The First-Person Pronoun

We will begin by contrasting 'I am *F*' with the more general 'Something is *F*'. One way of explicating the function of sentences of the latter sort is suggested by the following formula:

> 'Something is *F*' is used in English to express the following property of its utterer: that of accepting the proposition that something is *F*.

Let us note that the letter '*F*' is here schematic and may be replaced by any predicative expression. And 'is used to express' is an intentional locution. I think it is obvious that the relevant concept is one that is essential to the theory of language. In place of 'is used to express', we could say, more accurately, 'has as one of its uses that of expressing'. (One could abbreviate the above formula by saying that 'Something is *F*' is used in English to express the proposition that something is *F*. But the longer formula is preferable in that, as we shall see, it is applicable in situations where what is expressed cannot be said to be a proposition.)

'But I can say "Something is red" without meaning to say something about myself.' And this, of course, is true. But our formula does not say that one uses it in order to *say something* about oneself. The latter concept, as well as that expressed by 'makes the statement that', should be explicated in terms of our undefined locution.

We may now return to the first-person pronoun. I suggest that 'I am wise' expresses the speaker's property of being an x such that x directly attributes to x the property of being wise; in other words, the property of being an x such that x believes himself to be wise. Hence, we may say more generally:

> 'I am F' is used in English to express the following property of its utterer: that of being an x such that x directly attributes to x the property of being F.

It should be noted that the variable 'x' in the final clause is not a free variable. But we could also replace the final clause by: 'that of believing himself to be F'.

We have noted that the various types of believing may be reduced to direct attribution. But it does not follow from this fact that, in the ordinary sense of the words, all our beliefs are 'beliefs about ourselves'. For, in the ordinary sense of these words, I may be said to have a belief 'about myself' only if I directly attribute to myself some non-universal property; I don't, in this sense, have a belief 'about myself' if I directly attribute to myself the property of being such that all men are mortal. 'Saying' and 'talking about' are similar. If I utter the words 'I am wise' then I am saying something about myself. But if I utter the words 'Someone is wise' then, in the ordinary sense of the words, I cannot be charged with 'talking about myself'—even though, according to what has been said, I have directly attributed to myself the property of being such that someone is mortal. We have thus explicated the locution 'I am F' without appeal to 'I'-propositions.

'But', one may object, 'there are two difficulties with what you have said. (a) What is expressed by the sentence "I am F" may be true or false. How can this be if the sentence does not express a proposition? And (b) *what* one says, when one says "I am F", certainly stands in logical relations to various propositions. For example, the locution "I am *F*" logically implies "Someone is *F*"—and presumably the latter statement does express a proposition. What is the nature of this logical relationship?'

Let us consider the two objections.

(a) So far as the truth—or falsity—of what is expressed by 'I am *F*' is concerned, we may say simply this:

> 'I am *F*' is used *with truth* in English if and only if its utterer is *F*.

We must distinguish truth and sincerity in this context:

> 'I am *F*' is used *with sincerity* in English if and only if the property it is used in English to express is one that its utterer believes himself to have.

(We may note, in passing, that if the present account is accurate, the statement 'I am lying' may not be interpreted as 'The *proposition* I am asserting is false'.)

(b) And what of the *logical relations* between what is expressed by 'I am *F*' and what is expressed by 'Someone is *F*'? Given what we have said about the truth of 'I am *F*', we see that it is impossible for 'I am *F*' to be used with truth unless there is something that is *F*. And so one could say that the locution 'I am *F*' implies the locution 'Something is *F*'.

And so we may say, as Anscombe does, that 'if *X* makes assertions with "I" as subject, then those assertions will be true if and only if the predicates used thus assertively are true of *X*'. But where she has attempted to explicate the indirect or emphatic reflexive in terms of the use of the first person pronoun, the present account, if it is adequate, is an explication of the use of the first-person pronoun in terms of that intentional attitude that I have called 'direct attribution'—the attitude that we express by means of the indirect reflexives.

Notes

1 G. E. M. Anscombe, 'The first person', in *Mind and Language* ed. S. Guttenplan (Clarendon Press, Oxford, 1975), pp. 45–66.

2 Compare H. Castaneda, 'He: A study in the logic of self-consciousness', *Ratio*, 8 (1966), 130–157.
3 G. E. M. Anscombe, *op. cit.* p. 61.
4 H. Castaneda, *Thinking and Doing: The Philosophical Foundations of Institutions* (Reidel, Dordrecht, 1975), p. 159.
5 R. M. Chisholm, *Person and Object: A Metaphysical Study* (Allen & Unwin, London, 1976), Chapter 1.
6 R. M. Chisholm, 'The self and the world', paper presented to the 1977 Wittgenstein Colloquium in Kirschberg am Wechsel.
7 Compare G. Frege, 'The thought: A logical inquiry', *Mind*, LXV (1956), 289–311, and E. Husserl, *Logical Investigations* (Routledge & Kegan Paul, London, 1970), pp. 315–316. I defended this view in Chapter 1 of *Person and Object: A Metaphysical Study* (see Ref. 5 above).

[5] IDENTITY AND THE FIRST PERSON

Harold Noonan

> '. . . the mind knows itself, even when it seeks for itself, as we have already shown. But nothing is at all rightly said to be known while its substance is unknown. And therefore, when the mind knows itself, it knows its own substance; and when it is certain about itself it is certain about its own substance. But it is certain about itself, as those things which are said above prove convincingly; although it is not at all certain whether it is air, or fire, or body, or some function of body. Therefore it is not any of these. And to that whole which is bidden to know itself belongs this, that it is certain that it is not any of those things of which it is uncertain, and is certain that it is that only, which only it is certain that it is.' (St Augustine, *De Trinitate*, Book X, Chapter 10)

A person has self-knowledge, in the sense relevant to this paper, if and only if he knows some proposition to be true of the form 'I am *F*'.[1] He has identifying self-knowledge if and only if he knows some proposition to be true of the form 'I am *F* and only I am *F*'. It is tempting to suppose that self-knowledge entails identifying self-knowledge: that if one knows the truth of *any* proposition of the form 'I am *F*' one must know the truth of *some* proposition of the form 'I am *F* and only I am *F*'. This is tempting because of the plausibility of the view that 'I' is a singular referring expression. I shall first of all be arguing—or perhaps more accurately, giving a sketch of a possible argument—that if this is so, and if one's self-knowledge can be merely that one thinks and whatever else one *necessarily* knows in knowing this, i.e. the Cartesian primary certainties, then one's identity cannot be that of a human being—in Cartesian terminology, one must be really distinct from, not identical with, any human being.

If I have identifying knowledge of an object I can give it a proper name. But to give an object a proper name I need to know the appropriate criterion of identity for the type of object it is, since the introduction of a proper name requires its association with a criterion of identity.[2] And to know the appropriate criterion of identity for an object *X* is to know something about the changes *X* can undergo: it is to know that (at any rate) *X* will always be the same *such-and-such*—where 'such-and-such' is a term which conveys the appropriate criterion of identity for *X*.

Thus, when I have identifying *self*-knowledge I must know some proposition to be true to the effect that so long as I exist I shall always be the same such-and-such—where 'such-and-such' is a term conveying the appropriate criterion of identity for the type of object I am. Hence, if I can have identifying self-knowledge while knowing only that I think and whatever else I must *necessarily* know in knowing this, then it follows that even in this condition I must know some proposition to the effect that so long as I exist I shall always be the same such-and-such—where 'such-and-such' conveys the appropriate criterion of identity for the type of object I am.

But in knowing only that I think and whatever else I necessarily know in knowing this I do *not* know that I am a human being; I do not even know that if I am a human being I shall always be the same human being so long as I exist. These things may be true, they may even be necessarily true. But it is possible to doubt them while being certain that one thinks.[3] Thus the proposition to the effect that so long as I exist I shall always be the same such-and-such which—if self-knowledge entails identifying self-knowledge—I must necessarily know in knowing that I think, cannot just be the proposition that so long as I exist I shall always be the same human being. Nor can this latter be something I have to know in knowing the former. Again, it cannot even be something I have to know in knowing this that *if* I am a human being I shall always be the same human being so long as I exist. Yet, if a human being is what I essentially am, in the sense that the appropriate criterion of identity for me *is* that for a human being, then the proposition to the effect that so long as I exist I shall always be the same such-and-such, which—if self-knowledge entails identifying self-knowledge—I must necessarily know in knowing that I think, and which I can know without knowing either that so long as I exist I shall always be the same human being, or even that if I am a human being I shall always be the same human being so long as I exist—this proposition must nevertheless number both the latter propositions among its entailments.

If there is *no* proposition to the effect that so long as I exist I shall always be the same such-and-such, which I have to know when I know myself to think, and which satisfies the other conditions indicated just now: then if self-knowledge entails identifying self-knowledge, and if I can possess self-knowledge while knowing only that I think and whatever else I *have* to know in knowing this, it follows that the appropriate criterion of identity for the object I am is not that for a human being—that is, that I am really distinct from, not identical with, any human being.

Now the question whether such a proposition exists may be rephrased as follows to make it easier to consider. Knowledge of the appropriate criterion of identity for an object is arguably something very like knowledge of what Aristotelian (or Augustinian) second substance it is: as I have already indicated in passing, it can reasonably be described as knowledge of the *kind* of thing it *essentially* is. Our question can therefore be rephrased in this way: given that I am essentially a human being, is there any way I can latch on to the kind of thing I essentially am while knowing only that I think and whatever else I must necessarily know in knowing this?

Now as far as I can see there is only one way to defend an affirmative answer to this question, and that is to say that I may latch on to the kind of thing I essentially am by an *indexical* description; just as, in certain circumstances, I may latch on to the kind of animal I perceive, not as 'a bear' or 'a horse' or whatever, but merely as 'the kind of animal over there'. In the same way, it might be said, even if I am essentially a human being, I can latch on to the kind of thing I essentially am, merely as 'the kind of *thinking thing* here'; and in order to do this I need not know anything more about myself than that I *am* a thinking thing—or equivalently that I am thinking—just as I need not know anything more about the animal I perceive than that it *is* an animal in order to latch on to the kind of animal it is by the description 'the kind of animal over there'.

The suggestion can be further developed—it was so to me in conversation[4]—by saying that as *animal* stands to *bear* as genus to species, so *thinking thing* likewise stands to *human being*. Human beings, of course, are also a species of animal, but being a species of *rational* animal, they are also a species of *thinking thing*, which is a genus including (perhaps) not only (some) animals but also such things as God and the angels (each angel counting as one species of the genus).[5]

It is crucial to this suggestion as stated, of course, that just as one can perceive animals by outer sense, so one can perceive thinking things (at least one thinking thing) by a sort of 'inner sense'. Or at any rate, unless the suggestion involves this it becomes unintelligible: for then we are given no handle on how to interpret the description 'the kind of thinking thing here'.

Another version of the suggestion does not involve this consequence. It might be said that it is not as 'the kind of thinking thing here' that I can latch on to what I essentially am, but rather as 'the kind of thinking thing thinking this thought', as I may latch on to a kind of animal I can hear but not see as 'the kind of animal making that noise'. This version of the suggestion might be preferred by someone with Humean doubts about the observability of the self. However it still requires the existence of an 'inner sense' by which I may at least perceive my thoughts. For otherwise the interpretations of the descriptions 'the kind of animal making that noise' and 'the kind of thinking thing thinking this thought' must come apart, and then our understanding of the former gives no aid in understanding the latter. Thus on either version of the suggestion the existence of an 'inner sense' is a requirement of its intelligibility.

Now consider propositions of the form 'I am thinking so-and-so'. One cannot be mistaken in saying that one is thinking so-and-so, and one cannot be thinking so-and-so without being able to say it (although one may not be able to say it out loud). More specifically, in saying that I am thinking so-and-so I cannot be mistaken about the *identity* of the thinker: I cannot have taken someone else to be myself and because I know him to be thinking so-and-so have mistakenly concluded that *I* am thinking so-and-so.

To accommodate these facts the suggestion under consideration has to be developed, in its first version, as follows. It has to be maintained that the 'inner sense' by which I can perceive myself *qua* thinking thing is necessarily restricted in its scope to myself. For suppose it were not: then I could latch on to a kind of thinking thing as 'the kind of thinking thing here' without necessarily latching on to the kind of thinking thing I myself essentially was. Secondly, I could use the description 'the kind of thinking thing here' and fail to latch on to any single kind of thinking thing at all. Thirdly, I could misidentify another thinking thing as myself and so take myself to be thinking something which I was not thinking. For similar reasons the second version of the suggestion has to be developed by maintaining that it is a peculiarity of inner sense that an individual thinker's inner sense is necessarily restricted in its range to *his own* thoughts.

These developments would not be needed to meet the first two points I have raised (though they would still be needed to meet the third) if it were possible to maintain that there was only *one* kind of thinking thing. But remember where we are: the suggestion under discussion is being considered as a way of defending a position, a component of which is that I am essentially a human being (the other components being that self-knowledge entails identifying self-knowledge and that I can possess self-knowledge while knowing only that I think and whatever else I must necessarily know in knowing this). In this context, to maintain that there was only one kind of thinking thing would be tantamount to saying that human beings were the *only* kind of thinking thing, and this seems scarcely plausible. Even in the form[6] that the appropriate criterion of identity for any thinking thing is that for a human being it still seems implausible. Moreover, it could not have been accepted for a moment by anyone with the theological commitments of Descartes or St Augustine—and this is not merely an irrelevant historical aside since, as I expect is obvious by now, what I am trying to do is to develop in more modern dress what I take to be the dominant—or, at any rate, the most promising—line of thought underlying their dualism.

Setting aside then the possibility of maintaining that human beings are the *only* kind of thinking thing, the suggestion under discussion depends for its intelligibility on the thesis that I possess an inner sense which is necessarily restricted in its range *either* to myself *or* to my thoughts. But this dependence, I think, is fatal to the suggestion, for how might such a necessary restriction in range be understood? Consider the question with reference to the second version of the suggestion.

It is tempting to think that the reason that one cannot know the thoughts of another in the same way as he does is that there is a sort of barrier opaque to one's inner sense, behind which his thoughts lie. But we cannot really take this picture seriously. What cannot be seen because it lies behind a barrier could be seen if the barrier were down, but we don't want to allow the possibility of any circumstances in which we could know the thoughts of another in the way that he knows them.

Another idea that comes to mind is that we should think of the impossibility of knowing another's thoughts in the way that he does by analogy with the impossibility of seeing sounds or hearing colours. According to this idea, the reason why my inner sense cannot be extended in its range beyond my own thoughts is that my own thoughts constitute the entire class of its proper objects. But this entails that there are as many *kinds* of inner sense as there are individual thinkers, and that my thoughts, your thoughts, and someone else's thoughts have no more in common than a colour, a sound, and a smell. Why then are they all rightly called 'thoughts'? And how can this be so if, as is being maintained, human beings are *one kind* of thinking thing? How can that description be correct if the thoughts of different human beings are no more alike than are sounds and colours, and if their means of access to them are no more alike than hearing and sight? But if we set this second idea aside how else might the necessary restriction of my inner sense to my own thoughts be understood?

So far as I can see only one possibility remains, which was suggested to me by John Locke's discussion of personal identity. According to Locke, if one *thinking substance* could remember doing that individual act which was done by another *thinking substance* this would not entail that one *person* could remember doing that individual act which was done by another *person.* On the contrary, if one thinking substance did remember doing what was done by another thinking substance that would merely make the two thinking substances *one* person. Locke was thinking of diachronic identity only, but his idea may be applied to the case of synchronic identity too. One might maintain that *this* is the reason why one person cannot perceive by inner sense the thoughts of another: any other thinking thing the thoughts of which any given thinking thing has access to by inner sense thereby counts as the *same person* as the given thinking thing—whether or not it is the same in any other way.

However, this proposal can help to explain how my inner sense is necessarily restricted in its range to *my own* thoughts only if *my own thoughts = the thoughts of whatever is the same person as me*. But this is so only if what I am essentially is a *person*. So the proposal does not help to explain how my inner sense can be necessarily restricted in its range to my own thoughts if what I am essentially is a human being—which is the relevant question: all it comes to is a stipulation that if I perceive by inner sense the thoughts of another human being then he may be called the same person as me. But this plainly gets us nowhere.

None of the proposals, therefore, seems to give a satisfactory account of the idea that my inner sense is necessarily restricted in its range to my own thoughts. But this idea is crucial to the second version of the suggestion under consideration as to how one might defend an affirmative answer to the question: given that I am essentially a human being, is there any way I can latch on to the kind of thing I essentially am while knowing only that I think and whatever else I must necessarily know in knowing this? This line of defence of the possibility of an affirmative answer to this question must therefore be abandoned. And of course the same goes for the other line of defence, associated with the first version of the suggestion, since precisely analogous difficulties are involved in the idea of a necessary restriction in the range of my inner sense to my own self. But since these seem to exhaust the possibilities of defending an affirmative answer to our question, as I remarked earlier, it appears to follow that I must either deny that I am essentially a human being, or deny that self-knowledge entails identifying self-knowledge, or deny that there is any such minimal state of self-knowledge as knowing only that I think and whatever else I must *necessarily* know in knowing this. While I prefer the last of these courses it is, of course, more important to see what the options are than to decide among them.

Nevertheless, supposing that I am essentially a human being a further important conclusion follows from the argument so far. This is that 'I' may be used correctly *when it is not being used to make reference to any object*. For I might be in a state of 'sensory deprivation': blind, paralyzed, unable to touch any part of my body with any other; moreover I might be suffering from amnesia so that I do not know who I am or even that I am a human being. Yet I could still ask myself such questions as 'Who am I?', 'What am I?', 'How did I get into this state?'[7] In these questions, however, 'I' could not have a referring role, since reference is only possible against the background of a criterion of identity but, as we have seen, in such a state a man has no means of latching on to the criterion of identity for the type of object he is.

I think that the distinction between using 'I' against the background of a criterion of identity, to refer to an object, and using it *not* in this way is what Wittgenstein in the *Blue Book* calls the distinction between the use of 'I' 'as object' and the use of 'I' 'as subject'.[8] One uses 'I' (or 'my') 'as object', Wittgenstein says, in such sentences as 'My arm is broken', 'I have grown six inches', 'I have a bump on my forehead', 'The wind blows my hair about'. One uses 'I' 'as subject' in such sentences as '*I* see so-and-so', '*I* hear so-and-so', '*I* try to lift my arm', '*I* think it will rain', '*I* have toothache'. 'One can point to the difference between these two categories', Wittgenstein continues,

> by saying: The cases of the first category involve the recognition of a particular person, and there is in these cases the possibility of an error . . . The possibility of an error has been provided for . . . It is possible that, say in an accident, I should feel a pain in my arm, see a broken arm at my side, and think it is mine, when really it is my neighbour's. And I could, looking into a mirror, mistake a bump on his forehead for one on mine. On the other hand, there is no question of recognizing a person when I say I have toothache. To ask "are you sure that it's *you* who have pains?" would be nonsensical . . . this way of stating our idea suggests itself: that it is as impossible that in making the statement "I have toothache" I should have mistaken another person for myself, as it is to moan with pain by mistake, having mistaken someone else for me. To say "I have pain" is no more to make a statement *about* a particular person than moaning is.[9]

But if the interpretation of Wittgenstein's distinction which I have suggested is correct, what is the link between my utterances of 'I'-as-object and my utterances of 'I'-as-subject? It cannot be that they are all cases of *referring* to a particular individual. Is it merely that when I use 'I' 'as subject' the mouth from which the 'I'-token issues is the mouth of that very man who is the reference when I use 'I' 'as object'? But could this not be imagined otherwise? In her essay 'The first person' Professor Anscombe invites us to imagine someone standing before us and saying 'Try to believe this: when I say "I", that does not mean this human being who is making the noise. I am someone else who has borrowed this human being to speak through him'. And, as she says, our imagination is capable of making something of this idea: but then it is capable, also, of making something of the idea that in using 'I' 'as object' I might not be referring to that human being from whose mouth come the sounds when I use 'I' 'as subject'.

In order to get clearer about the link between the use of 'I' 'as object' and the use of 'I' 'as subject' let us first consider whether the use of 'I' 'as object' could exist on its own, or whether it needs to be coupled with the use of 'I' 'as subject'.

Of course, one could employ different words for the two categories of use (say 'I' and 'Ego'), so that no single word was used in both ways, but that is not what we are concerned with.

Nor, exactly, are we concerned with the question whether there could exist uses of the word 'I' 'as object' with *no* word having its characteristic use 'as subject'. For the answer to this question is plainly affirmative too. (In inflected languages the use of the first-person pronoun is optional; we might speak such a language but regard it as obligatory to employ the first-person pronoun where its use would fall in the category of 'use as object' and as forbidden otherwise.)

Our question is whether the use of 'I' 'as object' would be possible for someone who lacked the *capacity* to use it 'as subject'—where this capacity is something that is manifested not only by our ordinary use of 'I' but would also be manifested by speaking in either of the two ways just mentioned.

Now at first sight, I think, it looks as if this question should be answered affirmatively too. An argument for this is suggested by Professor Anscombe's discussion in 'The first person' of the society of '*A*'-users. (She describes the '*A*'-users as not having self-consciousness but as employing '*A*' as a name which everyone has but no one uses except to refer to himself.) She writes:

> Imagine a society in which everyone is labelled with two names. One appears on their backs and at the top of their chests, and these names, which their bearers cannot see, are various: "*B*" to "*Z*" let us say. The

> other, "*A*", is stamped on the inside of their wrists, and is the same for everyone. In making reports on people's actions everyone uses the names on their chests or backs if he can see these names or is used to seeing them. Everyone also learns to respond to utterance of the name on his own chest and back in the sort of way and circumstances in which we tend to respond to utterance of our names. Reports on one's own actions, which one gives straight off from observation, are made using the name on the wrist. Such reports are made, not on the basis of observation alone, but also on that of inference and testimony or other information. *B*, for example, derives conclusions expressed by sentences with "*A*" as subject, from other people's statements using "*B*" as subject . . . Thus for each person there is one person of whom he has characteristically limited and also characteristically privileged views: except in mirrors he never sees the whole person, and can only get rather special views of what he does see. Some of these are specially good, others specially bad. Of course, a man *B* may sometimes make a mistake through seeing the name "*A*" on the wrist of another, and not realizing it is the wrist of a man whose other name is after all not inaccessible to *B* in the special way his own name ("*B*") is.

That '*A*', in this society, is, as Professor Anscombe claims, a name which is the same for everyone but used by each person only to refer to himself (that is to say, there may be occasions when *B*, say, calls '*A*' someone other than *B*, but such uses of '*A*' will be *mistakes, misuses* of '*A*') seems correct. For it is evidently Professor Anscombe's intention that the sense of '*A*' in *B*'s mouth should be such that *B* can correctly call '*A*' only something: (i) of which he gets the special views she refers to; *and* (ii) of which the sense of '*B*' is a mode of presentation.[10] But now let us imagine a variation on the fantasy.

Consider why Harold Noonan's calling Smith 'I' is wrong. Is it because Harold Noonan is then calling someone other than Harold Noonan 'I'? No. For Harold Noonan might not know that he is Harold Noonan and still be able to use 'I' correctly. Harold Noonan has *learnt* to respond to utterances of the name 'Harold Noonan' in the appropriate manner, but it is no part of his understanding of 'I' that he recognize that in his mouth it is equivalent to 'Harold Noonan'.

Let us imagine then that it is no part of the understanding of '*A*' that is required of *B* by his fellows in this society that he assent to '*A* is *B*'; what is required if *B* is to show a grasp of the sense of '*A*' is merely that he recognize that he must only call '*A*' *that* item of which he gets the characteristically limited and also characteristically privileged views Professor Anscombe speaks of. Now that item is *B* himself. So it looks as if '*A*', in our variation on Professor Anscombe's fantasy, is an expression which is necessarily used correctly only if used by each person to refer to himself, while yet being such that for any member of the society, where '*N*' is that person's name, '*A* is *N*' may be something he does not know. Since the members of this society lack the use of 'I' 'as subject', does this not seem very like a description of the use of 'I' 'as object' in separation from the use of 'I' 'as subject'?

However, this can be maintained only if it really is the case that, for example *B* would *necessarily* be using '*A*' incorrectly if he were to call *C* '*A*'. But this is not so. For that *B* is the object of which *B* gets a special view is not a necessary fact, but a consequence of the nature of *B*'s perceptual apparatus. If this were tampered with so that the view *B* previously got of *B* he now got of *C*—and this is especially easy to imagine if we think, as Professor Anscombe suggests, of the '*A*'-users as machines rather than people—then there would be *no* misuse in *B*'s calling *C* '*A*': in so using it, in fact, he would be using it in just the way he was obliged to in order to use it correctly, given its sense (we can of course speak of the sense '*A*' has in someone's mouth, as we cannot, if I am right in my interpretation of the notion, speak of the sense 'I' has when used 'as subject').[11] This becomes obvious once it is realized that there would be no way of explaining to *B* the mistake he was making with '*A*'. Before *B* was tampered with, if he asserted '*F*(*A*)' as a result of misidentifying someone else as '*A*', one could show him that that person was not the one of which he got special views, hence not the one the sense of '*A*', as used by him, was a mode of presentation of. But after *B* has been tampered with, if he asserts '*F*(*A*)' as a result of observing that '*F*(*C*)' what can one say to him? To tell him that *B* is not *F* will not help since he no longer believes that he is *B*, i.e. he no longer assents to '*A* is *B*'. To tell him 'You are not *C*' will not help since if he checks on this (i.e., checks on whether or not he should assent to '*A* is *C*'), he will find that he should. And there is just nothing else to say to him. Philosophers like McTaggart, attempting to assimilate the use of 'I' 'as subject' to its use 'as object', sometimes suggest that so used it is equivalent to 'this self'. But their idea is not that *any* legitimate use of 'this self' is a legitimate use of 'I': rather it is that 'I' is correctly used if it is used in the way 'this self' *would* be used if it were used by any self only to refer to that self presented to it in inner perception.[12] They then face the task of explaining why it is impossible for any self to have an inner perception of any other self—since, according to their account of the meaning of 'I', any self a self has inner perception of, whether or not it is itself, is something it can correctly call 'I'.[13] The defender of the view that '*A*' is equivalent to 'I' in its use 'as object' faces the analogous task of explaining why it is impossible for *C* to be so presented to *B* that *B* would be correct in calling it '*A*'—except that in this case it clearly *is* possible and the task is evidently hopeless.

'*A*', then, is not necessarily used incorrectly by a member of the society we have imagined if one man calls another '*A*'. But 'I' in its use 'as object' *is* used incorrectly if used by anyone except to refer to himself. So the use of '*A*' is not the use of 'I' 'as object' divorced from the use of 'I' 'as subject'.

A consequence of the fact that I can only use 'I' to refer to myself is that I must always use it to refer to the same thing. In this respect the use of 'I' 'as object' is unlike the use of '*A*' and like that of an ordinary proper name.

But in another respect the use of 'I' 'as object' is like the use of '*A*' and unlike that of a proper name. In order to determine whether a mistake has been made in using 'I' 'as object' or in using '*A*' no recourse is needed to what is the case at times other than the time of utterance of 'I' or '*A*'. In Wittgenstein's example, when I mistakenly say 'I have a broken leg', my mistake can be established by determining that neither of the legs I had at the time of utterance was broken. Similarly if *B* should say '*A* has a broken leg' he can be proved to be mistaken by showing that the body of which he was capable of a special view at the time of this utterance had no broken leg. By contrast, since the reference of a proper name is that object which at a certain time fulfilled a certain condition, to establish a mistake in the use of a proper name *may* require recourse to what is the case at times other than the time of utterance of the name.

It is the way in which the use of 'I' 'as object' is thus partially like and partially unlike the use of a proper name that makes for the crucial difficulty in understanding how it could exist in separation from the (possible) use of 'I' 'as subject'.

But even *given* the use of 'I' 'as subject', one might ask, how can such a thing exist as we have seen the use of 'I' 'as object' to be? How can there exist a use of an expression which is in this way partly like and partly unlike the use of a proper name? After his perceptual apparatus had been tampered with *B* could correctly call *C* '*A*'. Indeed, given the sense of '*A*' in his mouth, he had to use '*A*' to speak of *C* if he wanted to use it correctly. In its use 'as object' 'I', in my mouth, has a sense too, since it then has a reference and there can be no reference without sense. But then how is it that the only object of which that sense is ever a mode of presentation is myself—Harold Noonan? That would be intelligible if the sense could be expressed in a description of the form 'the object which at time *t* satisfies condition *c*', but if that were so the use of 'I' 'as object' would not be unlike the use of a proper name in the way that it is.

Our problem is to explain why it is an impossibility for a man *A* to be using 'I' correctly when he is using it to speak of another man *B*. Now if this were to occur two things would be the case: first of all there would be a voice issuing from the mouth of *A* which was using 'I' correctly and yet using it to speak of *B*. And secondly that voice would not merely be issuing from *A*'s mouth but would, in addition, be *A*'s voice—because it would be *A* and not, say, some spirit possessing *A* who was speaking.

For the voice to be using 'I' correctly in speaking of *B*, however, it would have to stand in the same relation to *B* as we stand in to ourselves. When Harold Noonan is visibly in pain I am able to say sincerely, without observation, 'I am in pain'; conversely when I am able to say sincerely 'I am in pain' then *mostly* Harold Noonan is visibly in pain (only 'mostly' of course since one can hide a pain). When Harold Noonan's body is moving in a certain way or is disposed in a certain arrangement I am mostly able to say sincerely, without observation, 'I am moving thus-and-so' or 'I am disposed thus-and-so'; and when I am able to say sincerely 'I am moving thus-and-so' or 'I am disposed thus-and-so' then mostly Harold Noonan's body is moving thus-and-so or is disposed thus-and-so.

When Harold Noonan sees a tree or hears a yell than I am able to say sincerely, without observation (of Harold Noonan) 'I see a tree' or 'I hear a yell', conversely when I am able to say sincerely 'I see a tree' or 'I hear a yell' then mostly Harold Noonan is seeing a tree or hearing a yell.

If the voice issuing from *A*'s mouth is to be using 'I' correctly in speaking of *B* then there must be a similar correlation, unmediated by observation, between its pronouncements and the behaviour and situation of *B*'s body. That is to say, unless there is such a correlation between those of its utterances in which 'I' is used 'as subject' and the behaviour and situation of *B* it will be inappropriate to describe it as using 'I' correctly when it uses it 'as object' to refer to *B*. For in the absence of such a correlation it will be inappropriate to describe the body of *B* as the body of the one who is speaking when the voice speaks, but 'I' in its use 'as object' has the force of 'my body'.[14]

But if it is to be *A* who is speaking when the voice speaks, if the voice's assertions are to be *A*'s assertions, then a similar correlation must exist between the voice's pronouncements and the behaviour and situation of *A*. If *B* is in pain and *A* is enjoying himself, then, what should the voice be willing to say? If *B* is standing and *A* is sitting what should the voice be willing to say? Whatever the voice says, so long as it restricts itself to the use of 'I', either a correlation between the pronouncements and the behaviour of *just one* of *A* and *B* will be established, or an imperfect correlation will be established with both of them. Either way there will not be a clear-cut case of *A* using 'I' correctly but to speak of *B*. Of course, the voice may not restrict itself to the use of 'I': it might for example say such things as 'I-*A* am in pain and I-*B* am happy' or 'I-here am sitting and I-there am standing'. Such an idea could perhaps be elaborated in such a way as to make it plausible to say that *A* was using, say, 'I-there' to refer to *B*. But what to say of such a case is unclear, since against describing it as a case in which *A* is correctly using 'I' to refer to *B* stands the fact that if the sense of 'I-there' is the sense of 'I' in its use 'as object' then the sense of 'I-here' cannot be the sense of 'I' in its use 'as object', since the senses of 'I-there' and 'I-here' are different—but 'I-here' seems as good a candidate for 'I' as 'I-there'. Perhaps the most significant point is that in trying to think of a situation in which someone uses 'I' correctly but to speak of someone other than himself we have ended by considering a case in which someone uses two expressions in place of the single expression 'I', and in such a way that in the case of *one* of these ('I-here') he *would* be using it incorrectly if he were to use it to speak of anyone other than himself.

Notes

1 Thus, not every item of knowledge one has of oneself is self-knowledge in my sense, since one may know something to be true of oneself under the name or description '*X*' without knowing that one could truly assert 'I am *X*'.
2 This, of course, is something I am just assuming—which is one reason why I said my 'argument' was perhaps more accurately described merely as an argument sketch. But I think the assumption is widely enough accepted for it to be worthwhile to see what can be based on it.
3 This is obvious for the first of these propositions; as for the second—recall John Locke's views on personal identity.
4 With Mary Geach.
5 This development of the suggestion, actually, is not just an optional extra. 'The kind of animal here' means more than 'the kind of thing to which belongs something here which is an animal' (contrast 'the kind of red thing here') otherwise it could never secure reference. Similarly, 'the kind of thinking thing here' must mean more than 'the kind of thing to which belongs something here which is a thinking thing'; but to make this out would be to go a good way towards showing that *thinking thing* is to *human being* as genus to species.
6 Which is slightly weaker: the appropriate criterion of identity for an object can be that for an *F* even if the object is not an *F*.
7 See G. E. M. Anscombe 'The first person', in *Mind and Language*, ed. S. Guttenplan (Clarendon Press, Oxford, 1975).
8 But the matter is difficult. The clearest distinction is between using 'I' knowing the kind of thing one is and using 'I' not knowing this. Then there is the distinction between those sentences containing 'I' or 'my' which may be asserted on the basis of an identification of someone as oneself and those which may not. And finally there is the distinction between those occasions when one asserts a sentence containing 'I' or 'my' on the basis of such an identification and those when one does so not on such a basis. This last distinction is needed as well as the second one, since there are sentences like 'My arm is moving', which may be, but need not be, asserted on the basis of the identification of someone as oneself. I take it that 'I' is used 'as subject' whenever it is not used on the basis of such an identification.

9 L. Wittgenstein, *The Blue and Brown Books* (Basil Blackwell, Oxford, 1969), pp. 66–67.

10 This explains why she thinks it obvious that the use of '*A*' involves *re*-identification.

11 I once saw a film of the assassination of Trotsky, in which Trotsky's viewpoint was that of the camera: this may serve to indicate the sort of 'tampering' with *B*'s perceptual apparatus that I have in mind.

12 As Professor Anscombe points out, the intelligible use of a demonstrative pronoun does not require the presence of a referent but only the presence of something for the utterance to latch on to: I may say 'these ashes' meaning the ones in the urn, though I cannot see the ashes, but only the urn. The ashes, if they exist, are my referent, but what my utterance latches on to is the urn. In the same way, I could use 'this self', in such a statement as 'this self is getting annoyed', to refer to the self connected with a particular body, even if that self was not an object of perception for me, so long as its body was. In the use of 'this self' which philosophers like McTaggart imagine to be equivalent to the use of 'I', however, the referent and what the utterance latches on to have to be identical, i.e., the referent has to be present to consciousness—and present, moreover, in a certain special way.

13 McTaggart, of course, thought that selves *could* have inner perceptions of other selves. I do not know whether he had any way of blocking the apparent consequence of his views that a self could assert 'I am *F*' and be right even when it was *not F*.

14 But, of course, 'my body' does not mean 'that which will be a corpse after I die'; my body = the living human animal that I am.

II

ACTION AND INTENTION

[6] ACTION AS EXPRESSION
Charles Taylor

The natural expression of wanting is trying to get.

I'd like to try to explore what a claim of this kind can add up to. In what sense can action be seen as expression? I want to tackle this problem in two stages. In the first, I discuss the different senses that 'expression' could have. In the second, I look at the question whether action can be seen as expression in one of these senses.

Part I

What do we mean by expression? I'd like to suggest that we can articulate what this term means in a wide range of its uses with the formula: an expression makes something manifest in an embodiment. Both the key terms, 'manifestation' and 'embodiment', point to necessary conditions. I can make something manifest by clearing away the obstacles to its being seen. I can draw the curtain and make the painting plain for all to see. But this kind of manifestation has nothing to do with expression. On the contrary, my face expresses joy, my words express my thoughts or feelings, this piece of music expresses sadness, longing, resoluteness. In all these cases, something is made manifest; but what is manifest is so in what I want to call 'embodiments', and I want to claim that they are essential to what we call expression.

But the kind of manifestation we call expression is very special. It is not a sufficient condition of an expression that *X* be manifest in *Y*, where *X* is not identical to *Y*. For instance, seeing your car outside college, I know that you are in today. Your being here is, we might say, manifest to me in your car's being in the lot. But your car's being there isn't an expression of your presence. Expression involves a manifestation of a different sort; it involves what we might call direct manifestation, not leaning on an inference.

When I know something or something is plain to me,

through an inference, there is something else which I know or which is plain to me in a more direct way, and which I recognize as grounding my inference: in the above example, your car's being in the lot. It is characteristic of expression that it is not like this. I see the joy on your face, hear the sadness in the music. There is no set of properties that I notice from which I infer to your emotions or to the mood of the music. In those cases where I may have to infer how you feel (perhaps from a slight trembling of your hands which I've learned to recognize as signs of agitation or anger in an otherwise perfectly controlled 'front' you put up to the world)—in these cases, we cannot speak of expression. Your hand-tremble doesn't express your feelings, as your face might if you let yourself go.

An expression must at least offer what we might call a 'physiognomic reading'. I want to speak of this, in cases where we can see *X* in *Y*, where *X* is not identical to *Y*, and where there is not some other feature of *Y*, *F*, which permits us to infer to *X*. We have cases of physiognomic reading where, for instance, I look at a construction and see that it is highly unstable and will shortly fall. I read the impending fall in the upright array; but when challenged to say how I do it, I cannot cite features of the array on which I base my judgement which are only contingently linked with the impending fall. I may be able to point to particular sections of the array which reveal its imminent plight most clearly, but I can only characterize these in such terms as 'instability', i.e., in terms which are logically connected with what I see impending. Or again, I may be able to tell that this painting is by a given painter; and I may even be able to point to the features which are most characteristic; but I can't articulate what they have in common other than in terms like: 'a characteristic flourish of *X*'.

Expression must at least offer a physiognomic reading. But as can be seen from the above examples, this is not a sufficient condition. The unstable array doesn't express its impending fall. There are stronger conditions on expression. There are two I'd like to adduce here:

(i) In cases of genuine expression, what is expressed can only be manifest in the expression; whereas in mere physiognomic reading, the *X* we read in *Y* can be observed on its own. If I'm there at the right time, I can see the construction actually fall, or see the painter painting the canvas. Here I observe in a way free of physiognomic reading what I formerly read in the array or canvas. But the joy you now see in my face, the thoughts my words now make evident for you, the sadness or resolution in the face of fate that you hear in this music; all these could be manifest in no other sort of way. True, they might find some other expression: I might dance for joy; speak to you in another language; and resolution in the face of fate might be expressed in someone's demeanour. But these things can't be made manifest in a way which avoids expression, as the collapsing construction is observed free of any physiognomic reading.

(ii) The second difference between expression and what merely offers a physiognomic reading can be put intuitively in this way: that in a physiognomic reading, some *X* can be observed in *Y*, but in a case of genuine expression, the *Y* makes *X* manifest. The expressive object can be said to reveal what is expressed in a sense stronger than simply allowing it to be seen.

There are two ways in which this distinction can be made less rough: The first concerns our criteria for determining what is expressed/made evident. In the case of a mere physiognomic reading, some *X* can be read in *Y* just in case someone is capable of reliably discerning *X* non-inferentially in *Y*. The clumsy array offers a physiognomic reading of its impending fall just because you and I can *see* that it's about to collapse. What is offered here for such a reading is simply a matter of what we can reliably see.

But the criteria for determining what an expressive object expresses are somewhat stronger. It is not a sufficient condition of *Y*'s expressing *X* that someone be able reliably to see *X* non-inferentially in *Y*. I may know you well, and be able to see from your latest sketch that you were very tired or under stress when you did it. But this is not for all that what your work expresses. The work may by contrast be giving expression to a serene vision of things, or to a view of things as alive with energy.

The point here is that we see the expressive object not just as permitting us to see something but in a sense as saying something. The word 'saying' may not sound quite right in certain contexts, but the point is that with expressive objects, their expressing/saying/manifesting is something that they *do* in a sense, rather than something which can happen through them. And that is why what something expresses can not just be a matter of what an observer can read in it.

In other words, we can apply verbs of utterance, like 'saying', 'expressing', 'manifesting', to expressive objects; and this is more than just an optional metaphor; rather it is the indispensable background to a question which essentially arises with expressive objects: what are they saying? as distinct from, what can be read in them?

The important thing about an expressive object is that the verb of utterance applies properly to the object, and not just in a transferred way, as short hand for an agent's saying or manifesting something with the object. In other words, what *Y* expresses, if *Y* is an expressive object, or what *Y* makes manifest *qua* expression, is not reducible to what is made manifest by the emission of *Y*. In talking I make manifest the timbre of my voice, but this is not expressed in my voice. Or again, I can make something manifest by doing something which is not at all expressive action. I pull back the curtains to reveal the picture; or I arrange for some noise to direct your attention to that corner where you will see what I want you to see. In cases of genuine expression, the object must be said to manifest something, where this is to be attributed to the object, in a way that can't just be reduced to its emission or utterance manifesting that something.

This offers us another possible criterion for distinguishing expression from what offers a mere physiognomic reading. With genuine expressive objects, we can apply a whole range of what can be called adverbs of utterance. A work of art, or someone's words, or a facial expression can say what they say equivocally or unequivocally, uncertainly, in hints and fragments, confusedly, or gently, or strongly, or with conviction, or overpoweringly, or enigmatically, or obscurely, or hesitatingly, etc. And here again, the application of these can't be reduced to their application to the act of creating, emitting, uttering. That is why there can be an *issue* about how relevant an artist's intentions are to the understanding of his work.

The distinction I'm drawing here between expression in a

full-blooded sense and what only offers physiognomic reading seems to fall within the range of what we normally call facial expressions. I might describe the expression on your face now as 'tired'. But a tired expression is not expression in the strong sense. We mean nothing more by it than that it allows us to read your fatigue non-inferentially in your face. No further question can arise about whether your face really expresses fatigue.

By contrast, take the smiling expression on the face of someone who is open and welcoming. The smile expresses openness, welcome. But we are now saying more than that we can read welcome in it, as we read the tiredness above. The smile *communicates* openness, it doesn't just allow it to be seen. The smile plays an essential part in setting up communication between people, because it doesn't just allow openness to be seen, but communicates the disposition to be open, to communicate, to be in the other's company. In other terms, an essential part of what you see in a smile is the disposition to manifest, to communicate. What the agent is disposed to manifest, here openness, welcome, is thus expressed in the smile in a strong sense. I may also see in your smile that you are tired, but that concerns expression in a different, weaker sense.

There is a range of our facial expressions and stances which are expressive in this strong sense, because they are essentially concerned with the disposition to manifest our feelings: expressions of joy and sorrow, a stance of dignity, reserve, intimacy, etc. It is these that I wanted to take as paradigm examples, along with works of art and the words we utter, of what I am calling 'expressive objects'.

But is a smile an 'expressive object' in the meaning of the discussion above? Might we not say, contrary to what I claimed about works of art above, that the smile's expressing openness just reduces to the agent's manifesting his openness in smiling, as he might have done in a host of other ways? I would argue not. The smile's expressing friendliness is not just a matter of its being used by an agent to communicate his friendly intent. In the rough conditions of mutual marauding on the frontier, I may communicate my friendly disposition to you by refraining from attacking your compound; but this doesn't express friendliness like a smile. That is because the smile's link to manifesting openness is anterior to the framing of all intentions to communicate anything. The smile plays a crucial role ontogenetically in our being able to enter into communication as human beings in the first place. Its expressing friendliness cannot thus just be a matter of its being deployed with that intent; although of course a smile may go dead, become a grimace, or look sinister, if the intent is absent.

We might sum up the above discussion by saying that for a genuine expressive object, *Y*, its expressing *X* can't be reduced either to its offering a physiognomic reading of *X*, or to its being emitted/created/uttered with intent to communicate *X*. And we might add that it can't be reduced to the sum of these conditions either. If I contrive to make something manifest in a physiognomic reading, it still doesn't amount to expression. Thus I might work all night, so as to look tired, so as to get the point across to you that you ought to be nicer to me. But this still doesn't make my tired expression into an expression in the strong sense.

So to sum up this first part, I've tried to articulate the conditions that something must meet to be an expression. I've tried to come closer to an adequate account by stages. An expression manifests something, but in an embodiment; and not any kind of manifesting-in-embodiment will do, but one that offers a physiognomic reading; and not any kind of physiognomic reading will do, but one that is such that:

(1) what is manifest cannot be observed in any other way than a physiognomic reading; and
(2) some verb of utterance can be attributed to the object, and not just of the agent in emitting/creating the object, so that: (a) what it expresses is not just a matter of what can be physiognomically read in it; and (b) we can apply adverbs of utterance to it.

Part II

So what to make of the idea that our action is an expression of desire ('trying to get' an expression of 'wanting')?

There are some things we could think of which are unproblematically expressions of desire: I want some tasty food which I see over there: I mime, lick my chops, make by eyes like saucers, rub my stomach, say 'ummmm'. Or else, I can say, 'I'd love a piece of that delicious layer-cake'.

Another thing I might do is just go up to the plate, cut a piece of layer-cake, and eat it with relish. Is this also an expression of the desire?

We resist assimilating the action to the mime and verbal statement. This is because these two are clearly expressions in the strong sense. They manifest my desire, and this in a sense which is irreducible either to their just offering a physiognomic reading of the desire, or to their figuring in a successful attempt on my part to make my desire evident to you.

They are irreducible in the first way, because mime and statement arise as part of our activity of communicating, of being open with each other. Gestures and words are shaped by their role in communicating. And so there is a question of what they say which is not reducible to what can be read in them. The case of mime shows the contrast perhaps most strikingly; because it builds on certain natural reactions which do allow mere physiognomic readings of desire (or perhaps even in some cases, inferential readings). When I am very hungry and I see good food, my eyes get bigger, I salivate (making me lick my lips), my stomach rumbles, etc. These are natural reactions; the step from them to the exaggerated caricature is the step to communication; it has a place in the way of life of beings who have to be in a sense open with each other. It manifests the disposition to communicate.

And mime and statement are irreducible in the second way, because although closely bound up with the disposition to communicate, their manifesting my desire is not just a matter of my intentions in deploying them. I may even give myself away involuntarily, if I blurt out what I want or let one of these gestures of desire escape me. On the other hand, I could get you to be aware of my desire by causing you to be aware of my galvanic skin responses; but these would not for all that be expressions of desire.

So mime and statement are expressions in the strong sense. In the case of the mime, we may even be tempted to adopt the term 'natural expression'. Except that this might be misleading, since there is always an element of the conventional and arbitrary in all human communication. We have to find acceptable in our culture this kind of self-revelation in order to communicate effectively through it. It might seem grotesque or tasteless in another culture; and this to such a degree that it might even become hard to read; it might be mistaken for lewdness, for instance.

But there is still a motive for speaking of the mime as a natural expression. Because it builds on what I called the natural reactions, features that allow a physiognomic reading of hunger. We might easily by a natural extension of the term call these 'natural expressions' as well, as the reactions which are most easily taken up into expression in the strong sense. And this might suggest one construal of the formula that trying to get is the 'natural expression' of wanting. For among the 'natural reactions' of desire, our actions in trying to get occupy a central place. It would be quite normal, in our mime of desire for the cake, to act out reaching out and grabbing it, stuffing our mouths, etc. In fact a tremendously important range of genuine expressions build on what emerges to physiognomic reading in action. It is the range of what we might call bodily style. By 'bodily style', I mean what we refer to when we say of someone that he habitually acts cool, or eager, or reserved, or stand-offish, or with much sense of dignity, or like James Bond. The style consists in the way the person talks, walks, smokes, orders coffee, addresses strangers, speaks to women/men, etc. It is a matter of how we project ourselves, something we all do, although some do it in more obtrusive fashion than others.

There can be such a thing as bodily style, as a manner of self-projection, embedded in the way we act, just because so much of how we feel and react to things emerges to physiognomic reading in our actions. In our action can be seen our hesitancy, our reluctance, our eagerness, our fastidiousness, our ability to take it or leave it. Bodily style builds on this as the cake-desire mime builds on eyes opening and mouth salivating; it shapes the gesture in order to project, to make a presentation, of what the gesture offered as physiognomic reading. If I am semi-bored, not really engaged in any of my conversations, then my voice will lack spirit and animation, I shall speak slowly, without great emphasis; so projecting the bodily style of the cool, self-sufficient, uninvolved person will involve stylizing these traits of the voice into a drawl. If I am enthusiastic, interested in things, anxious to meet people and glad to communicate with them, then my voice will be animated, my manner quick to respond; and this too will be stylized into the 'eager' manner, when I come to project myself.

If we wanted to define 'natural expression' as 'the natural reactions which offer the most accessible physiognomic readings which can be taken up into genuine expression in mime or style', then the actions of trying to get would certainly be the most important and central natural expressions of our desires, because so much of the quality of our motivation can be read in our action, as we saw above.

This is one sense that can be given to the formula I'm trying to construe. It would respect the reluctance we saw above to assimilate trying to get to other expressions of desire like mime and statement which are full-blooded expressions. It would involve using 'expression' in a weak sense, which involved merely offering a physiognomic reading. But it would restrict the term 'natural expression' to a specially privileged locus of physiognomic reading, in that what it offers is taken up into expression in the strong sense. In this sense, it is clear that a (perhaps the central) natural expression of wanting is trying to get.

I think that this is true. But it is rather unexciting. 'Expression' only figures in this construal in an attenuated sense; and it is not clear that action has a specially privileged position as natural expression. I should like to push for a richer construal.

A richer construal seems possible because some analogue of the two conditions mentioned in the previous section, which distinguished full expression, also seem to hold of action as an expression of desire.

The first point which distinguished genuine expression from what merely offers physiognomic reading was that with the latter what we read can also be observed outside of a physiognomic reading. We can see the wall falling on its own, as it were, as well as reading its impending collapse in its present tilt. But what is expressed can only be manifest to us in an expression.

In this regard, the desire–action relation is more like genuine expression. Our desires can be manifest in our actions; they can be manifest in other features of our demeanour which offer a physiognomic reading of them; and they can of course be manifest in genuine expressions of them, like the mime or declaration above. But they cannot be manifest altogether outside a context of expression or physiognomic reading. It is not even conceivable what it would be like for them to be so independently manifest. Desire is not a candidate for manifestation outside some medium.

Part of the difference here can be put in this way: in the case of the unstable construction, there are quite clear spatio-temporal considerations allowing us to distinguish what is read from the context in which it is read. In this case, we have two different temporal phases of the same collection of objects, at one moment in unstable array—where their impending collapse can be seen—at the next in the process of collapsing. But this kind of separation can't be made in the action–desire case. The desires that we predicate of agents are not discriminable spatio-temporally from the person who acts, in whose behaviour we read the desires.

So much is true, of course, of any physiognomic reading we make of persons—even of those cases where we see that, for example, they are tired, or depressed—i.e., where we are not discerning desires through action. But there is a special inseparability point to be made in the action–desire case, and this is the well-known (perhaps notorious) one, that our desires are characterized by the actions they dispose us to undertake.

This is the point which has been articulated in the logical connection argument, as it has sometimes been called, that it is essential to the notion of desire, that if we want *X*, have the ability, and the opportunity, and no restraining rival motive, we will do *X*. The thesis is that some formulation of this conditional can be taken as an *a priori* truth, determining what will count as a desire. The significance of this point has been widely misunderstood. It doesn't raise an issue for whether desires can be causes of the actions they explain. Plainly they are. Rather the point is that desires can't be identified without reference to the actions they tend to produce. And this connection offers the basis for seeing actions as 'natural expressions' of desire in a stronger sense.

It provides in fact an analogue for the second important difference between genuine expression and what merely offers physiognomic reading. This was that the question what an expression manifests cannot simply be reduced to what can be reliably read in it non-inferentially.

Now something analogous seems to hold when it comes to reading desire in our action. Of course, a great deal about our motivation can be physiognomically read in our action. You may see by the way that I am pumping the water, or picking the cotton, that I am dreaming of Mary Lou, and longing for her return. If you know me well enough, you can make a great many quite penetrating readings of this kind from my behaviour. In the weak attenuated sense, my longing for Mary Lou is expressed in my behaviour.

But there is a stronger sense in which my desire for *X* is expressed in my trying to encompass *X*. This is the sense that my action here displays the identifying characteristics of my desire. Now the issue of what a given action expresses in this stronger sense doesn't reduce to what can be veridically non-inferentially read in it. That was indeed sufficient for the weaker sense of expression invoked in the previous paragraph. My cotton-picking expressed my longing for Mary Lou just in case someone could discern in my action this state of my soul. But in the stronger sense of this paragraph not all the soul-states discernible in the action are expressed in it, but only the desire whose defining characteristics are displayed.

So we have an analogue of the second feature of strong expression in the previous section. Action manifests desire in such a way that there is a question of what it manifests that does not reduce to what can be seen in it. And this stronger sense of expression applies specifically to action, and not to those other natural reactions, for example, eyes opening wide, trembling, etc., which also offer physiognomic readings of our feelings and motives.

It might be thought that the question of what desire is expressed in an action in this stronger sense is just the question of which of the desires physiognomically evident in it is also its cause. It is true that the desire here causes the action. But it is not true that we will be able to understand this relation just with the aid of some general category of causal relation.

Because in some general sense of cause, there can be a desire which can be read in our behaviour and which causes it, and which nevertheless is not expressed in the behaviour. This is so in cases of what have been called 'deviant causal chains': for example, the case where I desire very much to smash your Ming vase because it will pay you out for being so mean to me. This powerful destructive desire of mine makes me so nervous that my grip is loosened on the vase and it smashes on the floor.

This kind of case has been much discussed in the literature. And of course there is no agreement on what to make of it. My claim is that it shows that the relation of action and desire cannot be analyzed with some general category of cause which is also applicable to inanimate beings. Rather we can see that just what is missing in the deviant case is that relation between desire and action that I want to call 'expression'. Someone can physiognomically read my hostility to you in the agitation which eventuates in my dropping the vase; but this doesn't express my desire in the stronger sense, even though by fluke the resulting event (the smashing of the vase) turns out to have the same description as the one I wanted to bring about. We can see from the deviant case that this stronger, more intimate relation of expression is essential to what we consider normal action. What is this relation? It is that desire and action are in a certain sense inseparable.

To see this we have to look at what one might call the normal or basic situation, one in which I act unreluctantly and unconstrainedly. This situation can be called 'basic', because here my action and desire are in the relation from which my classification of desires is drawn. My desires are always characterized by reference to this situation, where I am doing what I want. And this is the only way that a desire can be characterized. If I didn't have this normal situation as one of my possibilities, and at least sometimes as an actuality, I couldn't have the language of desire that I have.

To say that desire and action are inseparable in this situation is not to say that they can't be cause and effect. On the contrary, they are. But it does mean that they cannot be identified as separable components in this situation. My desiring X is not something separable from my unconstrained, unreluctant action encompassing X. And my action encompassing X

cannot be construed as behaviour which is brought about by desire as some separably identifiable antecedent. It must be experienced as qualitatively different from reluctant action, or non-action.

Let's look more closely at these two claims. First, the claim that desiring *X* is not something separable in the normal situation from the unconstrained, unreluctant pursuit of *X*. This would not be true if we could construe desire as some underlying neurophysiological state which issued in this behaviour, but this we cannot do. Desire is an intentional state. It is essentially bound up with a certain awareness of what is desired, of its object.

Of course, there are unconscious desires. But what do we mean by these? We mean desires of which our awareness is distorted. But even these, our hidden or self-unavowable desires, are manifested in affect of some kind. It is just that this affect quite misrepresents or distorts or screens what we consider the real nature of the desire.

I am not just making the point that the concept of a hidden desire is parasitic on that of a desire of whose object we are aware; that we couldn't attribute hidden desires to beings who didn't have conscious desires. This is true; but I'm making the stronger point that we couldn't attribute a hidden desire where this had no manifestation at all in conscious affect—although we may include among conscious affects certain pathological states of indifference, the inability to feel anything, in situations where feeling something is normal. What was utterly unmanifested in affect couldn't be a *desire*. (And it is difficult to see how it could even play the explanatory role of desire: something that makes me do certain things. For I must feel *something* when I do them; and how can *this* be unconnected from the desire which is allegedly issuing in these deeds? even if what I feel is a strange kind of numbness and distance from the actions?)

So we can't understand unconscious desire on the model of some unobservable physiological condition which produces external observable consequences, for example, infection with the scarlet fever virus, which produces the rash. In that case the connection is contingent between inner condition and visible symptoms; the condition could be there without there being any visible symptoms at all. But in the case of desire, there must be some manifest affect. It isn't a desire otherwise.

Now if we take this point that desire is an intentional state, and look back at what I have called the basic situation, we can see how desire and action are inseparable. For in the case where I act unconstrainedly and unreluctantly, my awareness of my desire must always include, and can just consist in, my awareness that I am so acting. But since my awareness of my desire is not some separable symptom, but is essential to my desire, is part of what it is to have this desire, then so is my awareness of my unconstrained action. Awareness of what I want is inseparable here from awareness of what I am doing. Desire and action are not separable components in the basic situation.

They begin to come apart when I am constrained from action. Then the awareness of desire can take the form merely of a formulation to myself, or to you, of what I want; or a sense of unease, perhaps. But this case is parasitic on the normal case, in that our desires are characterized by reference to the normal case.

So there must be a context in which my desire can't be separated from my action. But if this is so, then the second claim must also be true, that my action encompassing my end can't be understood as distinct from my behaviour in other contexts just in that it has some special separably identifiable antecedent. The awareness which enters into, which can be identical with, my awareness of desire, must be of a kind of action qualitatively different from other contexts. It is happy action.

It is because there is and must be such a context where desire and action aren't separable, this context being central to the development of our language of desire and action, that seeing action as the natural expression of desire is saying something stronger than just that action offers a physiognomic reading of our motivations. It does offer physiognomic reading of many desires, or rather, my total behaviour, demeanour, etc., does; as it also offers reading of my other states: fatigue, distress, nervousness, etc. But in the normal situation, there is a desire which enjoys a special status, which is the one inseparable from this happy action. To say that this is naturally expressed in the action is to say more than just that the latter offers a physiognomic reading of it.

Rather the claim is that the desire is in the action in a stronger sense. For the desire is inseparable from the action in that the awareness which it essentially involves just is the awareness of unconstrained action. So that in the case of happy action, we can say that the form the desire takes is that of unconstrained action. The locus of desiring in this case, as an essentially intentional state, is just in the action. The action doesn't just enable us to see the desire; it *is* the desire, embodied in public space.

Moreover, this is not just one among many forms the desire can take. It is in a sense its paradigm form, since the desire is defined in terms of the action which expresses it in this sense. Desires are defined in terms of the actions which can embody them in this way.

Thus action expresses desire in a stronger sense. It doesn't just make it visible, as it does my fatigue or nervousness, for instance. Rather happy action is the desire embodied in public space, in an action from which the desire is inseparable, and which therefore displays its defining characteristics.

We thus have a stronger construal to give our formula 'the natural expression of wanting is trying to get' than our earlier one which put it on all fours with formulae like 'the natural expression of fatigue is drooping shoulders', or '...of hunger is salivation', etc. And this relation of expression is essential to action. It is at its most palpable in what I have called the basic situation, where we act unconstrainedly and unreluctantly; but it is present in all action, since even when we act reluctantly or under constraint, our action is expressive of our intention or resolve in the same way that happy action expresses desire; that is, our awareness of our intention incorporates, and may be nothing more than, our awareness of what we intentionally are doing.

This is what underlies the distinction between deviant and non-deviant chains. In the normal, non-deviant case, our action expresses our desire or intention. This relation is a primitive, because it involves inseparability of the two. Consequently, it is futile to try to give an account of what non-deviancy amounts to in terms of a type of causal relation between separably identifiable terms. This might serve as a neurophysiological reductive account of what underlies the distinction; but if we want to talk about what the distinction is about in our experience, what underlies the intuitive distinction between deviant and non-deviant chains, this kind of independent-term-causal account just *must* be barking up the wrong tree.

The point is missed because in the objectivist orientation of much contemporary philosophy, we allow ourselves to construe desire as some underlying state, not necessarily intentional. It is something which may issue in (perhaps distorted) consciousness or avowal, as well as in action. But the underlying state is seen as independent from its results, in thought or action. On this view, the relation of action and desire in the basic situation isn't even seen as a possible option. Happy action is not seen as a special case, where desire and action interpenetrate; but rather just as a situation marked by an absence of conflict, and perhaps also an absence of other symptoms of desire than action, for example introspective awareness of longing, etc. On this view, action is no more intimately linked with desire than any of the other symptoms.

From this perspective, desire becomes the cause of action in a normal, Humean way; and it makes sense to try to find what special conditions distinguish deviant from non-deviant cases. But this way of talking is a muddle. It won't allow for the necessary intentionality of desire. It is an attempt to give an account on the psychological level, with terms like 'desire' and 'action', while operating with a logic that would only be applicable on the reduced, physiological level. The result is confusion.

At the centre of our ordinary understanding of action is thus an expressive relation stronger than what is involved in our action and demeanour offering mere physiognomic readings of our feelings and condition. Trying to get is indeed the *natural* expression of desire, not just as offering the most readily available physiognomic access, but also as what is inseparable from it by 'nature', i.e., the fundamental facts about the human condition which are determinative for our language. Unconstrained, unreluctant action is the paradigm form of desire; the form of its self-awareness which is remembered in all other forms; its natural outlet.

But is it a natural *expression* of desire? Nothing in the above discussion closes the gap with what I called genuine expression, where the object can be understood to say something. In that sense, what we say in mime and speech is always expression in a sense which mere unconstrained action can't match. In that sense our original reluctance to speak of expression here stands.

But although it is not genuine, full-blooded expression, there is a reason to speak of action as the natural expression of desire. It is not just that it manifests desire by embodying it in public space, which gives at least some minimal justification for using the term. It is also that this relation is in a sense foundational for genuine expression.

We have genuine expression, as we saw, when we can attribute verbs of utterance to expressive objects. But to get to this stage we have to be able to recognize the disposition to communicate as embodied in gestures or artefacts in public space, as for instance human beings can recognize and respond to a smile. Without this recognition, we would never have the predicament of mutual communication which all our expressive activity presupposes, in which alone verbs of utterance have sense. But in order to recognize reciprocally the disposition to communicate, we have to be able to 'read' each other, our dispositions and feelings have to be potentially open, in public space. Our desires have to be manifest to others, to the potential community.

This is a kind of manifestation which is foundational for genuine expression, in that it is presupposed by it. This is the 'natural' level of expression, on which genuine expression builds, always with some degree of the arbitrary and the conventional. Mime and style take this up and make a language in which we can say to each other, as it were, what we believe ourselves to be. But there would be nothing to take up, if we weren't already open, if our desires weren't embodied in public space, in what we do and try to do, in the natural background of self-revelation, which human expression endlessly elaborates.

[7] HOW THEORETICAL IS PRACTICAL REASON?

Anselm Winfried Müller

When I was studying with Elizabeth Anscombe during the 1960s, 'practical reasoning' was one of the many topics which came into my philosophical curriculum as a result of her exacting as well as extremely generous teaching. Intended as a token of gratitude and friendship, the following contribution will be devoted to a question which is raised but, I think, not answered by Anscombe's publications on the topic: what sort of thought is a practical thought? and, more specifically: can practical reasoning be understood as theoretical reasoning in the service of practice?

Part I

In *Intention*, we find an example of an argument about contingent matters which 'is not practical reasoning: it has not the form of a calculation what to do, though like any other piece of "theoretical" argument it could play a part in such a calculation'.[1] What does that form consist in? What would make an argument play a part in a practical calculation? Classroom examples apart, practical reasoning is reasoning 'with a view to action';[2] and 'whatever is described in the proposition that is the starting point of the argument must be wanted in order for the reasoning to lead to any action',[3] while the premises do *not* say *that* the reasoner wants such-and-such.[3]

The premises are assertions like:

They have Jersey cows in the Hereford market

or

Vitamin X is good for all men over 60
Pigs' tripes are full of vitamin X
I'm a man over 60
Here's some pigs' tripes

What is wanted by the man who reasons from these assertions? In the first case: a Jersey cow, or, perhaps more accurately, the possession of a Jersey cow; in the second: what is good for him. The premises of the second example unfold the connexion between what he primarily wants and one of the things he can do for its sake (viz., taking some of what's here).

As a canonical form of practical reasoning one might suggest

(1) *A* is a way of attaining *B*
(2) So I'll do *A*

Here (2) goes proxy for the conclusion (which is an action *A* of the reasoner). (1), which may summarize a chain of considerations connecting things to be done for the sake of other things, mentions what is primarily wanted (*B*) and states a way of attaining it (*A*) which can be implemented immediately (that is, without any further consideration how *it* be done).

By what feature is (1) the form of a practical premise? Perhaps by the fact that it does not necessitate the conclusion? But there will occasionally be things that I *must* do (and often things I must not do) if I want to achieve a certain end; and a premise that mentioned such an indispensable condition would necessitate the conclusion, as long as the end is my end. By itself, (1) has nothing practical about it; it yields, for example (theoretical) conclusions of the form:

P, or *A* is a way of attaining *B*

It is in virtue of the context, then, that a judgement is classified as a practical premise. 'This is a way to the farm' is likely to be a practical premise if the man who is told or says it wants to get to the farm; and this end of his, in turn, will be likely to be revealed in his taking the way in question. He may, of course, wish to keep away from the farm. Then, too, 'This is a way to the farm' could be a practical premise, leading the man to avoid 'this' direction.

Practical premises can be ambivalent in yet another way. The following story is told about one of the Rothschilds: He was looking for a good horse at the Frankfurt horse fair. One of the dealers offered a horse for sale, praising its worth: 'Why, if you leave Frankfurt at 6 in the morning on this horse, you will be in Königstein by 8!' 'Ah', said the Rothschild, 'I shan't buy it then: what, after all, should I want in Königstein at 8 in the morning?' In other words, the conditional form of a practical consideration is no guarantee that its role is that of (1).

Considerations concerning the role of a practical premise

lead Anscombe, in an unpublished paper, to admit an indication of what is primarily wanted into the formulation of a practical syllogism, but not as a premise proper: '. . . the end ought to be specified, but the specification of the end is not in the same position as a premise'.

Part II

Here we may pause to ask what is to be admitted in the role of a premise to a practical conclusion. In our canonical form, (1) corresponds to many of Aristotle's examples of practical premises only if we allow, in the place of '*B*', evaluative expressions, such as 'what is good for . . .', or expressions containing a word like 'should'.[4] Such examples appear not only to mention an end but also to specify something as an end. On the other hand, in special circumstances, something's being pleasant or good for one or an action one should perform may be taken as a reason against it. So why not include, in the structure of any piece of practical reasoning, a premise specifying the end in view of which the reasoning is practical? (Are not some of Aristotle's premises, like ποιητέον μοι ἀγαθόν in *De motu animalium* (7, 701 a 16 f), meant to provide such a specification?)

Perhaps it does not matter whether the specification of the end is called a premise. It is certainly not one whose form can be subsumed under (1). Since, on the other hand, the notion of practical reasoning seems to appeal to criteria of validity, it is reasonable to locate among the premises everything without which the practical conclusion does not follow. So let us ask whether the denial of a practical conclusion (as symbolized by (2)) is inconsistent with the acceptance of premises (as represented by (1)). If we identify the conclusion of practical reasoning with the action in which it results (*A*), denial (or rejection) of the conclusion will have to be a voluntary failure to do *A*. But there is no inconsistency between accepting that *A* is a way of attaining (a possible end) *B* and refraining from doing *A*.

Let me exclude a possible misunderstanding: there is no inconsistency here even if *B* is *my* end—simply because the premise does not present *A* as the only way of attaining *B*. I do *not* wish to deny that an action A can be said to be a valid practical conclusion from practical premises even though alternative actions would have served as ways to B; these alternative actions A_1, A_2, A_3, . . . would correspond to (the possibility of) premises connecting A_1, A_2, A_3, . . . with B. For the sake of my argument, however, I shall consider the case where the premises represent A as the only way (or a necessary condition) of attaining B.

Acceptance even of such premises is consistent with a refusal to do A, since B may not be accepted, i.e. intended, as an end. But 'premises' whose acceptance is consistent with a rejection of the 'conclusion' seem to be incomplete. So, one might say, the performance of A, the practical conclusion, does not *follow* as long as a specification of the end signalling its acceptance is not included in the premises.

Suppose, then, we formulate a further premise of the form:

(0) Let me attain B

in order to achieve validity for the practical reasoning towards (2). What does (0) express? It is not a statement to the effect that the reasoner has an end B. (0) seems to express a want rather than a thought—the (formation of an) intention to attain B.

Now a man who thinks that A is the only way of attaining B will indeed be inconsistent if he also wants B and does not do A; or, perhaps, his want will not be an intention to attain B in these circumstances. But should we therefore treat this want, like the judgement that A is the only way of attaining B, as a mental *premise*? May it not play a different role in practical reasoning? I said that its premises should include 'everything without which the practical conclusion does not follow'. But in a sense this was wrong. As the tortoise has taught us, we must not include in the premises of a theoretical argument 'everything without which the (theoretical) conclusion does not follow'. To avoid an infinity of suppressed premises, we must not treat 'If P then Q' as a suppressed premise in the argument 'P, hence Q' even though, in a sense, the conclusion does not follow without the truth that if P then Q.

Could wanting, or intending, a certain end be considered as a *principle of inference*—on the following analogy: for someone to infer 'Q' from 'P' he has to believe that if P then Q; for someone to 'infer' action A from (1) he has to want B? The analogy is not convincing: First, for the theoretical reasoning to be correct, the belief concerning the entailment must be true; there is no comparable condition on the primary wants of practical reasoners (whence the 'principles of inference' can here vary from person to person and from time to time). Secondly, if a piece of practical reasoning can be assessed as valid or invalid, an assessment of it as valid will be based on a judgement about the logical relations between its premises and its conclusion; and it is this *judgement* which has a *prima facie* claim to being the analogue of a judgement (or implicit belief) that one's premises entail the conclusion of one's theoretical argument.

Are there any more plausible ways of looking upon the role of (0)? I do not know of any—apart from treating it as a kind of premise after all. The attraction of this option is a neat mirror image effect: theoretical reasoning, starting from the consideration that if *A* is realized *B* is, needs a further premise relating to *A* in order to arrive at a conclusion concerning *B*; whereas practical reasoning, starting from the same consideration, needs a further premise relating to *B* in order to arrive at a conclusion concerning *A*. But this view of practical premises must be supplemented by answers to two obvious questions: by what kind of logic can there be a reasoning from (0) and (1) to (2)? and, once more, what is expressed by (0)—if it is a premise but not a statement?

Part III

The following position seems to solve these problems: Practical reasoning terminating in an action is more or less like theoretical reasoning terminating in a judgement or a belief. But the (mental) performances or states of the reasoner which are involved in a theoretical inference are not what we consider when the logical relations between premises and conclusion are in question: A belief follows from beliefs not in the sense that one *believing* follows from others; rather what is believed as a result of correct reasoning follows from what is believed as given. Similarly, we need not worry whether an action can be a conclusion and a want figure as a premise: the *logic* of practical reasoning relates to what is believed, what is wanted and what is done. Thus, '*X* does *A*' is implied by '*X* attains *B*' together with 'Doing *A* is the only way of attaining *B*'; and '*X* attains *B*' is implied by '*X* does *A*' together with 'Doing *A* is a way of attaining *B*'.

If the position I have sketched is a correct account of practical reasoning, there is no such thing as practical reason, or, rather it is (theoretical) reason in the service of practice. 'In the service of practice', because—or when—it expounds thoughts which: (a) relate to possible actions and things that can be attained by acting; and (b) are grounds of actions. But the thoughts themselves as well as their logical relations are the same, whether the reasoning is 'practical' or not.

On this view, two components have to be distinguished in (0): this premise, a 'fiat', in Kenny's terminology, has, on the one hand, a propositional content, or 'phrastic', accessible to considerations of ordinary (theoretical) logic, which is also the logic of 'practical' reasoning ('I shall attain *B*'); on the other hand, in virtue of its 'tropic' (or 'neustic') it expresses a want which has *B* (or, perhaps, my attaining *B*) for its object. In a similar manner, an action can be called a 'practical conclusion' only in so far as its conception or description by the agent provides us with a propositional content which is the concern of logic ('I am doing *A*').

So wants need not figure as premises, nor, properly speaking, actions as conclusions—any more than beliefs (in the sense of 'believings') play either role. In a practical inference, as symbolized by (0), (1) and (2), the business of reason is to identify, for a given proposition ('I shall attain *B*'), a second one which both presents the reasoner as acting in a certain way, and forms the subordinate clause in a true conditional of some sort with the first proposition as the main clause. So that is all 'practical' reason amounts to: reason whose achievements serve the reasoner's practice. However, the position that has emerged is less simple than it looks. The innocent word 'serve' needs explanation. What sort of service could be meant?

Part IV

When I reflect in order to find out how to attain a given end—and, finally, in order to attain it—that is not like sweating in order to lose heat as a result of evaporation. Nor is the difference just one of consciousness: I may be conscious not only that I am sweating but also what that is good for; still, I cannot be said to be led to perspire by the latter consideration. Practical thinking is, in this, rather like its typical result: acting with an intention. The action is done *with a view to* an end and *on account of* my insight that it helps to bring this end about. This also seems to be the way I engage in practical reasoning for the sake of an end of mine. (More often than not, actions are done for reasons without any prior 'engaging in practical reasoning'. But this does not prevent us from attributing to the agent thoughts that articulate those reasons. Their expression can be elicited, for example after the action, and function as a criterion of its purpose or character. So there is nothing fictitious about these 'dispositional' thoughts, and what I have to say about the goal-directed character of practical considerations is to apply to them, too.)

If practical thinking is not a case of natural teleology, does this mean that it is a means towards ends much like the means it is concerned to discover or to present to one? 'If I chop these small trees down, I can make a raft . . . .'—'Why are you going through these considerations?'—'I want to take these things across this river, and in order to be able to do it I am doing some thinking.' (Or 'If I am to . . . , I must consider how to. . .'.) However, my thinking 'in the service of practice' then seems to *presuppose* a separate insight into *its* usefulness, i.e. a thinking concerned with that thinking in the service of practice. We cannot escape this result if all thinking is 'theoretical'; if, that is, even in a practical context, our thinking is only judging (of appropriate means) and does not by itself—in virtue of the kind of thinking it is—stand in the service of practice; if the thought of its purpose is no part of it. But, of course, if, quite generally, thinking could be practical, i.e. for the sake of some goal, only on account of a thought which relates it to this goal, there would be no end to the chain of such thoughts presupposed by any practical thinking, or rather, there would be no practical reasoning and no reasoned practice.

Part V

Part of my claim, then, is that practical thinking is thinking both intrinsically for the sake of an end and about this end and ways to it. So, one might argue, a practical thought has two aspects: first, its function—somehow internal to it—of helping towards the attainment of some aim; second, its content, viz., that such-and-such end is attained by such-and-such means. Hence we need not revise the account of practical inference which emerged in Part III; this is concerned with contents only, and they are of the same nature whether the reasoning is theoretical or practical. However, this is not quite so. We cannot characterize a practical consideration as one which is for the sake of what it is about. An example will show that this is insufficient. Imagine, first, a prisoner passing his time by doing sums. Clearly, his calculations are not a case of practical reasoning, as they might be if he were conducting them, for example, with a view to some financial transaction. His end—to pass his time—does not come into the premises of his reasoning. But now suppose it does, in the following manner: in order to pass his time, he thinks of ways of passing one's time. Here, we could say, his thinking is geared to an end whose realizability is, at the same time, its content. Yet it is not practical reasoning, just because it is done as a pastime. It would be practical only if the prisoner reasoned somewhat like this: I must somehow pass my time. In my situation, taking exercise (or doing sums, or thinking (= theorizing!) about lots of pastimes) would be a way of passing the time. *So* . . . .

Practical thinking is about one's end *as* one's end. But this formulation does not explain anything. It conveniently conceals a problem, viz., what is the role of the end in practical thinking if thoughts that are both for its sake and about it (and means towards it), nevertheless need not be practical? Let us try the following answer: to reason practically is to consider which actions can promote one's end, *with a view to* realizing this end *by means of these* actions. There is no reason to have any confidence in this explanation. A counter-example to it would presumably have to be more artificial or more ingenious than the prisoner one. But its inadequacy seems clear. It fails to entail our observation that a practical thought is not, as theoretical thinking can be, an 'ordinary' means towards an end (deployed on the basis of a separate consideration of its usefulness) but involves consciousness of its own practical function as well as a judgement relating means to that end.

The teleology of a practical consideration (i.e., the way in which it is for the sake of what it is about) is different both from natural teleology (where the question 'Why are you doing that?' *does not apply*) and from ordinary intentional teleology (where the answer to that question expresses the practical consideration that *resulted* in the doing). The practical consideration 'behind' a practical consideration, one might say, is part of it. If I consider the possibilities of taking something across a river, in order to distract myself, my awareness (or the thought, if it occurs) that I am doing this to distract myself is in no way constitutive of what I am considering. If, on the other hand, I consider, with a view to acting, how such-and-such can be taken across, my consideration's being conducted with a view to this end must be treated as somehow internal to its own content. (One might even say that the distinction between the content and the employment of a thought is of limited validity only.)

(The feature of practical thinking which I have just tried to describe is, I think, what led Aristotle to attribute it to a faculty other than that of theoretical reason. Of course, he also connects the theoretical with necessity and the practical with contingency. But note a curious feature of his *λόγος ὁ ἕνεκά τινος* and similar concepts, like *πρακτικὸς νοῦς* and *διάνοια πρακτική*: they are applied to a kind of thinking which relates to ends not only by content but also by 'cause' (see *De Anima* III 9, 432 b 19–433 a 3). So we have *νοῦς ὁ ἕνεκά του λογιζόμενος* (*ibid.* 10, 433 a 14), which is distinguished from theory *τῷ τέλει* (14 f); *διάνοια* is practical when it occurs *ἕνεκά του* (*EN* VI 2, 1139 a 35 f; see *EE* I 5, 1216 b 16–19); and *νοῦς* gives orders *διὰ τὸ μέλλον* (*De Anima* II 10, 433 b 7 f; see 433 a 19). Very frequently, passages like these do not permit us to interpret the reference to ends exclusively either as specifying an object of thought or as characterizing its function. I suggest that this is not an unresolved ambiguity but an appropriate fusion of aspects.)

Part VI

If it is true that a practical thought cannot be captured by giving its content on the one hand, and its function on the other, how can it be identified at all?

This question turns our attention in a wrong direction: it raises the expectation of a highly sophisticated definition. But practical thoughts are 'identified', primarily, in contexts in which they both find expression and lead to practical conclusions. It is the way their expressions are connected with practical contexts which helps to fix the meanings of these expressions and which has to be invoked, also, when the role of such expressions (and, thereby, the concept of practical reasoning) is to be explained in a general manner. Problems arise when, in a philosophical analysis of practical reasoning, one tries to reduce expressions of practical thinking to a component expressing propositional contents and a component reflecting its practical function.

I have tried to point one such problem out, and this has led me to the observation that practical reasoning is not theoretical reasoning in the service of practice and that its teleology is somehow inseparable from its content. I am not at present trying to give a more positive account of practical reasoning by examining the relationships between its expressions and its practical context. Rather I wish to return to the question of verbalized practical inference. How do the reflexions of Parts IV and V bear on this?

Part VII

We should remind ourselves that in a formulation of premises and conclusions little hinges on the form of words that is employed and much on its 'logical multiplicity', i.e. on the difference it makes, or can make, to a whole context. This is clear when we consider, for example, possible answers to the question 'Why are you lying down?' *One and the same reason* is mentioned in each of the following replies:

(3) To be well (later)

(Contrast: 'To be well'=to stay well.)

(4) To get well

(5) I am lying down to get well

(6) I want to (hope to/must) get well

(7) I am not well

(8) Lying down is a way of getting well

(9) I believe (It occurred to me) that lying down is a way of getting well

Thus, philosophers who investigate what reasons really *are* have a choice between states and changes (of the agent) (3, 4), links of final causation (5), intentions, desires, hopes, needs etc. (6), negative facts, or statements (about the agent) (7), laws of efficient causation (8), and thoughts (dispositional and occurrent) (9). (To give a complete list of possibilities is not one of my ambitions.) (7) does not give a reason, or at least not the one under consideration, if it can be expanded into 'I am too weak to stand up'. What I have in mind is rather an utterance accounting for behaviour *A* by drawing attention to the present lack of *B*, whose presence *A* is believed to result in.

I agreed that it is wrong to mention wants—as opposed to what is wanted—in the premises of a practical syllogism. This was not to deny that a form of words like (6) can express a practical consideration. Rather I wish to exclude three misconceptions:

(a) In reasoning how, for example, to get well, I do not consider my wanting to get well as one of the facts of the situation like those mentioned in (7) and (8). I *can*, rather perversely, employ (6) in the same role as (7)—if I reason, for instance, 'I want to get well. Now I want to (must) get rid of this want. Taking such-and-such drugs is a way of getting rid of it. So. . . .' Or even: 'I want to get well. My psychotherapist tells me to do everything I want to do. Now I am resolved to do whatever my doctors tell me to do. . . .'—with the unexpected conclusion that I lie down! Of course, in these examples, 'want to (must)' and 'am resolved to', respectively, do occur in the role originally envisaged for (6).

(b) *I want to A* is not a reason for *A*-ing. For the same 'reason' would be given by 'I am *A*-ing in order to *A*' (compare (5) with (6))—which possibly puts an end to a chain of reasons but does not add a last link; or by '*A*-ing is a way of (a means to) *A*-ing' (compare (8) with (6))—which is false. Moreover, what is a reason for doing *A* would seem to be a reason also for wanting to do *A*; but 'I want to do *A* because I want to do *A*', which no doubt can mean something, does not mean such a reason. We do, admittedly, speak of '*A*-ing for its own sake'; so why not admit the kind of 'reason' I have just rejected? But this would be a verbal manoeuvre and would not affect the upshot of my argument: that my wanting to *A* is not, from the point of view of a practical consideration, *a fact about* *A*-ing on all fours with the fact that *A*-ing is a manner of (or a means to) *B*-ing.

(c) The idea that the wanting must be mentioned in the premises may, again, be a version of the idea that the agent's state of wanting *B* provides the reality (the hard fact) with which his reason for *A*-ing should be *identified.* But reasons do not stand in need of this or any other kind of identification with something 'real'. Two actions are done for one and the same reason if they are done with a view to one and the same end. A reason can be *presented* in many different ways; and one such way is exemplified by (6).

Part VIII

If (6) and (8) give one and the same reason, it may be said, we cannot have *both* as premises of one piece of practical reasoning. But is this true? Suppose I say '*Q*' and you ask me 'Why do you think so?' My answer will, or may, give one and the same reason for my judgement, whether I reply '*P*' or, rather, 'If *P*, *Q*' or 'Since *P*, *Q*'. All the same, we represent the reasoning by:

P
If *P*, *Q*
Therefore *Q*

By analogy, my answer to the question 'Why are you lying down?' might be given a canonical form: 'I am to get well, and lying down is a way of getting well'. So we were not wrong in supplementing (1) by (0) to reach (2). However, as I said, little hinges on the form of words. Instead of the addition of (0) we can imagine a special stress on, say, 'attaining' in (1). This would correspond to 'Since *P*, *Q*' in lieu of '*P*; (and) if *P*, *Q*'.

What I actually speak or write when I express actual practical considerations will depend very much on the context, which includes among other things the knowledge and beliefs I presuppose in the addressee about what I am doing (or am up to, or have done), about what my ends are, and about what I believe to be conducive to what. Does it follow that there is no general way of deciding what is and what is not part of the practical reasoning of someone who does *A* in order to attain *B*? No, for whatever occurs to him and whatever considerations he may be prompted to express, we attribute certain thoughts to him when we say that he does *A* in order to attain *B*. We attribute to him a conception of *B* as his end, the belief that *A* is conducive to *B*, and the thought of doing *A*; and we do not mention any superfluous premise when we say he reasons from (0) and (1) to (2).

But, by the argument of Parts IV and V, such a presentation of his practical inference is not a full presentation of the thinking in question. Nor can we say that *the* propositional content of sentences like (0) to (2), something neither theoretical nor practical, is what the inference operates on—for the content of a practical consideration cannot be separated from its practical function: the conception of this function is *part* of the consideration as well as constitutive of its teleology. What we might say is that practical considerations have an 'aspect' under which they present material for entailment relations.

If practical considerations are goal-directed in the way I have argued they must be seen as being, there is no reason to confine the teleological element to one of the premises. It is true that (0), understood as a fiat, expresses not only a want but also a thought (a conception of an end) which is not of the kind of a judgement; whereas (1) has the form of a statement. Also, (0), which does not, for example, express a *wish*, shows—or can be taken to show—that *practical* thinking is being represented. But both kinds of premises—identifications of ends as well as presentations of means—are for the sake of practical conclusions and thus in the service of those ends.

Part IX

Is the teleological character of practical reasoning of any relevance to its formal side? Of course not—if this means: whether my considerations affect the entailment relations which may obtain between propositional aspects of practical thoughts. The model of these relations is just a case of *modus ponens*:

If I do *A*, I attain *B*
I do *A*
Therefore I attain *B*

But practical reason provides us with reasons for *doing* things: Its conclusions are practical ones and for them to be validly drawn we need more, and in a way less, than entailment. Let me put forward some conjectures on this point.

A practical consideration relating to an end and to means to this end is, on my view, intrinsically 'in the service of' the attainment of the end. This implies that it results in an action conducive to this end when such an action is possible. Or, roughly: practical reason must be such that, failing impediments beyond its ken, what it is there for comes about. Here I leave open the question what might incapacitate an agent (in such a way that we should say: he did not do anything conducive to *B* but he was up to it). But so much is clear enough: when a given end leaves us a choice of equally acceptable means, *this* does not (*pace* Buridan's ass) count as interference, that is, as an impediment preventing the attainment of an end *B* without calling into question the teleology (and, thus, the truly practical character) of the considerations relating to *B*. So there seems to be a close link between what I call the intrinsic teleology of practical reasoning and the rationality of inferring non-necessary conditions as practical conclusions.

Similarly, the 'seriousness' of a practical consideration often requires that a practical conclusion is derived which does not by itself bring the end about but merely enhances the chances of its realization. In such situations the action which forms the practical conclusion may be a necessary condition of the end, or it may be an element of some sufficient condition of it; in either case, the agent's practical thoughts do *not* exhibit the *modus ponens* entailment mentioned above (it is not unconditionally the case that if he does *A* he attains *B*). The pattern is at best in the background, in that practical rationality demands that his practical conclusion be connected in a specific manner with some *A* of which he believes: if I do *A*, I attain *B*. Even this is not quite true. There need be no *A* of which I believe that it is more than probable or just possible that it would get me *B*, in order for me rationally to draw practical conclusions. (The nature of the end, and of the alternatives to it, will no doubt put conditions on this rationality.)

If, thus, the practicality of practical reason seems, as it were, to give it extra inferential powers, this practicality is, on the other hand, responsible for its 'defeasibility'. All the practical considerations of one agent relate to the shaping of one and the same life; *it* is the 'practice' which practical reason ultimately 'serves'. Hence a piece of practical reasoning can be defective, even though its practical conclusion is related to the premises in the right way, if practical premises have been neglected which concern a further end of the reasoner's and present it as incompatible with his practical conclusion. (Is this why Aristotle, at least in theory, considered *the* (over-all) end as the starting point of practical reasoning? It is hard to see how the end in this sense can be specified for a man (or for man) in such a way as to provide a starting point for actual practical thinking.)

Part X

These observations could be said to relate to inferential peculiarities of practical reasoning. The teleology of this type of reasoning also demands certain restrictions on its content. Some of these restrictions have been taken for granted. Practical considerations must (in the end) come up with such means to an end as consist in types of behaviour which are (immediately) open to the reasoner. (There will be less agreement on what types of things can count as ends.) Both ends and means must be (seen as) 'future contingents'. No thinking whose content was not thus restricted could be in the service of what it was about.

However, the restriction extends beyond this. Or, rather, when, instead of (1), we use:

If I do *A*, I shall attain *B*

as a way of representing practical considerations (as is common in writings on the topic), we must not forget that this conditional stands in need of a special kind of backing. 'If I carry a hat, I shall be following the criminal' might be part of a piece of practical reasoning, but hardly in the role of (1).

Let me just mention that the kind of backing that could make the If–then a 'practical' one is not an obvious matter. My own formulation '*A* is a way of attaining *B*' does not really fit all types of case, both on account of the considerations advanced in Part IX and because of the variety of teleological connexions generally. Not all of these involve (efficient) causation (and some involve it in the way exemplified by raising one's arm in order to activate certain nerves); the connexion may be one of part and whole (like putting the plates on in order to set the table, but also: pronouncing 'tough' in order to pronounce 't'); or one of genus and species, or of species and particular case; the end may be a result and not, strictly speaking, an effect of the means. . . .

Here is the context also in which to discuss the legitimacy of passing from

If I do *A*, I shall attain *B*

to

If I do *A* and A_1, I shall attain *B*

I shall not try to investigate under what conditions the second sentence could possibly be *inferred* from the first *in the course of a practical argument*. But the intrinsic teleology of practical thinking will have to contribute to the criteria. Idle thoughts cannot be part of a correct practical inference. May they not be part of a valid practical inference? But is it to be called valid if its conclusion (perhaps, for example, doing *A* and A_1) cannot be said to be for the sake of *B*?

The reflexions I have put forward in this and the last sections may suggest a reexamination of the relationship between practical and theoretical inference. (1)-type practical premises seem to differ from (theoretical) if–then statements in ways that are relevant to the form of valid practical reasoning. The assumption of a mirror image connexion between practical and theoretical reasoning must be qualified. And a comprehensive account of the soundness of a practical inference should draw on the requirements of the intrinsic teleology of practical reason.

Part XI

If we cannot speak of a practical thought as having a theoretical content, what shall we say of theoretical thoughts?

They, too, seem to have functions. Yet would it not be more than paradoxical to say of theoretical thoughts that there was only a theoretical aspect to *their* content?

But there is no problem here. I do not, indeed, have an account of what theoretical thinking is. On the other hand, my point about practical thinking is based on the consideration that it cannot be conceived of as: using concepts and judgements in order to (be able to) act in view of a given end, as a result of recognizing that this use will be helpful. But theoretical thinking, where it happens to be useful, can be thus conceived. Whether it is done for its own sake or for the sake of something else, that kind of teleology (as opposed to its tendency to be true, or valid) is not intrinsic to it.

Part XII

Suppose the following objection were raised against the conception of practical reason as intrinsically goal-directed thinking: When we attribute thoughts, considerations, wishes, intentions, decisions etc. to a man, the evidence on which we do this consists in his behaviour, including utterances. His mental states and events are not given to us in any other manner. Also, when we reflect on our own thoughts, wants etc., these do not appear to us to come, so to speak, in clear contours as though they were tailored to fit our categories of intellective and appetitive functions of the soul. Rather, these categories are—like any other concepts, only, perhaps, more obviously—instruments by which we bring division and order into the diffuseness of reality. Without such division and order we cannot reflect on and find orientation in reality. Concepts like *thinking* and *wanting* serve our communication about an otherwise indistinct complex of observable human *behaviour*. Hence it is up to us to define our categories as seems helpful. There is, then, no compelling reason to supplement human wanting with a kind of reasoning which is intrinsically practical.

Whether or not one accepts the approach on which this objection is based, the conclusion can—and, I think, must—be rejected. Even if concepts are somehow man-made, there are necessities about the connexions of concepts. As long as we employ words like 'thinking', 'wanting', 'for the sake of' etc. in the usual way in order to describe human realities, their employment follows certain rules. If my argument holds, these rules do not allow the realities describable in terms of expressions of the form 'N.N. wants to do *A* in order to attain *B*' to be captured also in a terminology of wanting, of theoretical thinking and, perhaps, of causation.

The concept of practical reasoning as intrinsically teleological thinking seems indispensable in an account of such realities. It is required by the fact that, in accordance with those rules of our language, practical thoughts must be classified as purposive but neither as cases of natural teleology nor as means conceived by other practical thoughts.

I cannot here consider the roots of this conceptual necessity, nor take up the possible suggestion that an alternative set of concepts could be deployed to cope with the articulation of human behaviour. And I have not talked about the problem that there appear to be thoughts right on the border between theoretical and practical considerations (a problem which, I think, is not specific to my approach to the topic). Given these and other lacunae, I find consolation in the thought that what I have said will be controversial enough by itself.

Notes

1 G. E. M. Anscombe, *Intention* (Basil Blackwell, Oxford, 1960), § 33, p. 60.
2 G. E. M. Anscombe, *ibid.* § 33, p. 59.
3 G. E. M. Anscombe, *ibid.* § 35, p. 66.
4 G. E. M. Anscombe, *ibid.* § 35, pp. 65–67.

[8] THE ARGUMENT FROM DESIGN
Robert Hambourger

The argument from design for God's existence is involved with important questions about the conditions under which it is reasonable to believe that a state of affairs was brought about intentionally. In this paper I shall offer a version of the argument and defend it, if not quite in the sense of trying to show conclusively that it succeeds, then, at least, in the sense of trying to show that it deserves to be taken seriously. In Part I, I shall present a number of objections to the argument that, for the most part, are quite well known and, I think, quite weighty. Most are descendants of objections to be found in the writings of David Hume.[1] Then, in Part II, I shall present the specific version of the argument I wish to offer here and, finally, in Part III, try to show that it does not succumb to the objections raised at the start.

Part I

At the beginning, I would like to distinguish two sorts of argument from design. Arguments of both sorts start with the fact that many natural phenomena look *as if* they might have been produced by design and try to show that such phenomena really *were* designed, and from this they hope that it will be concluded that the universe as a whole was produced by the intentional actions of a single being. But then, the arguments conclude, since the universe appears to be good, and since it would take an extraordinarily wise and powerful being to design and create it, the universe must have been created by a wise, powerful, and good being, and this is God.

Now I think it is clear that any argument of this type will face difficulties when it tries to conclude, from the claim that many natural phenomena were designed, that the universe as a whole was produced by the design of a single being. For it seems possible that the universe itself was not a product of design, even if many important parts of it were, and, further, even if we are willing to grant that the universe as a whole was designed, it still will not follow that it was created by the design of a single being.

This, however, does not seem to be the most serious objection facing the argument from design. For one thing, even if it is granted, the objection will not rob the argument entirely of its power. It would be a significant achievement to prove that certain natural phenomena were results of intentional action, even if one could not prove that the entire universe was created by a single being, and this achievement, by itself, would be enough to show that something was seriously wrong with the atheist's standard picture of the universe. Also, though, even if a demonstration that important portions of the universe were created by design would not actually entail the hypothesis that God exists, it would seem that, nonetheless, it would render that hypothesis more plausible than any competing one. In a contest with polytheism, monotheism is likely to prevail, and one who comes to hold that many natural phenomena were created intentionally probably will come to believe in God. This objection, then, does not seem to me to be a crucial one, and I shall not deal with it further in what follows. Instead, I shall be concerned with attempts by arguments from design to show that certain features of the natural world are products of design, and it is in the ways that they try to do this that the two sorts of argument I wish to distinguish here differ.

Arguments of the first of the sorts I want to discuss are true analogical arguments. One begins by pointing to ways in which certain natural phenomena resemble human artifacts. An animal's eye, for example, is much like a fine machine. And even where the likeness is not so direct, natural phenomena often share with artifacts what Cleanthes, in Hume's *Dialogues*, calls '[t]he curious adapting of means to ends'.[2] That is, in the case of numerous natural phenomena, as in the case of human artifacts, states of affairs which plausibly could be desired as ends are brought about by phenomena that themselves might reasonably have been intended as means to those ends.

Once such resemblances are noted, arguments of the sort I have in mind proceed straightforwardly by induction. Certain natural phenomena have features in common with human artifacts. However, in many cases, namely, in cases of artifacts, phenomena with these features have been discovered to result from the intentional actions of intelligent beings. Further, in no cases have phenomena with the features been discovered not to result from such a cause; for we have not discovered that natural phenomena with the features are *not* ultimately the result of God's design, and human artifacts all are designed. But then, by induction, we can conclude that all phenomena with the relevant features are products of design and, so, that many natural phenomena are results of intentional action.

This argument, however, I think, must fail. One important principle of inductive reasoning is that, in making inductive inferences, one should not extrapolate from cases of one sort to others that differ too widely from them. For example, if one has studied a great many horses but no other mammals with respect to a certain anatomical feature and has found that all mammals studied have had the feature, one still cannot properly conclude that all mammals and, thus, all dogs have the feature. For the difference between horses and dogs is too great. The point here is not that one can never make inductive inferences from one sort of case to another. If one has examined many animals of numerous mammalian species, though no dogs, with respect to a given feature, and if all mammals studied have had the feature, then it might well be proper to conclude that dogs have the feature, even though there are significant differences between a dog and any of the animals studied. For here the animals that have been studied vary as greatly among themselves as they do from dogs. But when all the cases one has examined for a property are alike in significant respects, and all differ in those respects from a new case that is being considered, then one can have little confidence that the new case will be like the others.

This, however, I think, is exactly the situation in which we find ourselves, if we attempt to infer—from the fact that human artifacts have been designed, and natural phenomena have not been proven not to result from design—that natural phenomena that resemble human artifacts have been designed. It is not that there are not clear analogies between artifacts and some natural phenomena. I think there are. But there are also important respects in which natural phenomena do not resemble artifacts. Consider, for example, the eye. One difference between an eye and a machine is the materials out of which the two are made. But I do not think this is the crucial difference. Suppose that a 'mad scientist' some day should construct eyes exactly like natural ones out of flesh and blood. This, I think, would make us no more inclined than we would be otherwise to think that our own eyes were produced by design, nor would we conclude that the analogy between natural phenomena and artifacts was finally close enough for the argument from design to go through.

What seems to me a more important difference is this. Human artifacts, even in cases of automated production, result quite *directly* from intentional actions. Our eyes, on the other hand, while we were developing in the womb, originated from genetically controlled processes that themselves had natural causes, and so on, back as far as we can determine. These processes might have been the results of design, but, if so, the design seems, so to speak, to have been woven into the fabric of nature. And, it would seem, a similar disanalogy can be found in all cases between human artifacts and those natural phenomena that look as if they were produced by design.

This difference between natural phenomena and human artifacts, then, which involves the very features of natural phenomena to which a proponent of the argument from design would be most likely to point to justify his belief that such phenomena, if designed, must have been designed by a divine being, is, I suspect, a sufficiently great difference to block analogical versions of the argument from design. At any rate, I shall not pursue the attempt to work out such a version here. There is, however, another sort of argument from design which is not an analogical argument; and though it also faces serious difficulties, in the end, I think, much can be said for it. Indeed, the version of the argument I shall offer later might be classified as one of this sort.

Arguments of this second sort begin with the claim that, in numerous cases, desirable features of the universe have been brought about by complex states of affairs whose occurrences might seem totally fortuitous, if they were not produced by design. The fact that conditions on earth, for example, were suitable for the development of life probably depended on precise details of the planet's composition and the positioning of its orbit about the sun, and the earth might well have been lifeless, if these had been even slightly different as the planet developed.

Again, basic features of the physical world depend on the fact that bodies of the sort that exist will interact as they do, given the laws of nature that hold. If either the bodies or the laws had been sufficiently unlike what they are, the universe probably would have been quite different and, very possibly, less interesting than it is. Consider, as an example, the forces that bind particles together into bodies and physical systems, for example, the forces binding atoms to form molecules or the force of gravitational attraction. If these forces had been much weaker than they are, matter could not easily have been formed into stable configurations, and the universe might have been little more than a system of particles in flux. On the other hand, if the forces were significantly stronger than they are, it would seem that things would have been overly stable, and discrete, changeable bodies might be at a premium.[3]

It seems, then, not to be a matter of course that the universe is as impressive a place as it is. In many cases, desirable features of the universe would not have come about, unless seemingly unconnected states of affairs had come together in the right sort of way. But, also, one would think, it could not simply have been an accident that, in so many cases, things came together in ways that had such impressive results. Cases of this kind need explanation. However, someone who presents an argument of the sort I am describing would argue that such cases could not be explained unless they were results of design or, at any rate, that they could be better explained in this way than in any other. And if this is true, it would seem that we can conclude that in many noteworthy cases features of the universe were created by design.

Arguments of this second sort, notice, are not analogical arguments. They do not claim that the natural phenomena they hold to have resulted from design are very much like human artifacts. Instead, they hope to show on other grounds that the phenomena need explanation but can only be explained properly as results of design. Thus, these arguments seem to sidestep the problems that beset analogical versions of the argument from design. Nonetheless, arguments of this new sort are open to serious objections, and before presenting my own version of the argument from design I shall mention three that I think are the most serious:

(i) Arguments from design of the second sort depend on the view that certain natural phenomena have features that make it appropriate to explain them as results of intentional action, quite apart from any analogies that hold between them and phenomena that have previously been discovered to be products of design. But this would seem untrue. When we first come by the family of notions that are connected with intending, we are not, it seems, taught logical criteria that allow us to determine whether a state of affairs was brought about intentionally. Instead, we begin by learning to recognize cases of intentional *behaviour*, and, once we know what it is for a person to act intentionally, it would seem that we come to learn that states of affairs of certain sorts are brought about intentionally, only because we frequently find such states to result from actions we recognize as intentional. That is, it seems that there are no inherent features of a state of affairs that show it to have been produced by design. Rather, we can only know a state to have been brought about intentionally either by knowing directly that it was produced by intentional actions or by knowing it to be sufficiently like states that we have discovered to result from such actions. But if this is true, then the only sorts of argument from design that can succeed are analogical arguments, and we have already seen reasons to think that analogical arguments fail.

(ii) We can reach this same conclusion from more general considerations about causation. A justly celebrated feature of Hume's theory of causation is the thesis that there are no *a priori* connections between cause and effect.[4] One cannot by reason alone discover the cause of any state of affairs. One can only do this by having observed similar states to have been preceded by a given sort of occurrence in repeated instances. But then it will follow, again, that we can only know a state of affairs to have been caused by intentional actions if it is sufficiently like states we have already discovered to have had such causes, and if, as seems likely, the most general features of the universe are too dissimilar from those occurrences whose causes we have discovered for us to be able to reason by analogy from one to the other, then we will have to remain in ignorance about the ultimate causes in nature. These two objections, together with the objection against analogical arguments from design I presented earlier, seem to me to constitute the most serious challenge to the argument from design, and it should be noted that they challenge far more than just this single argument. Indeed, they call into question whether reason ever could provide adequate grounds for believing in the existence of God. For suppose one had adequate reason to believe in God. Then one would thereby have adequate reason to believe that the universe was caused by design. But if either objections (i) or (ii) succeed, one could not have such reason unless one had discovered states of affairs that both sufficiently closely resemble the universe as a whole and are known to have been produced by design. And, given the difficulties we saw with analogical arguments from design, it would seem that one could not do this.

(iii) Finally, non-analogical versions of the argument from design ask us to conclude that certain phenomena were produced by design, simply because no other adequate explanations seem to be available. However, even apart from the difficulties above, one should be suspicious of arguments of this form. It might, after all, be that alternative explanations are available but that we just have not been able to think of them. Indeed, it would seem that alternative explanations of many suggestive natural phenomena are available by using the sorts of devices employed by the theory of evolution in biology. For that theory seems to show a way in which purely natural processes can result in the most highly organized and impressive sorts of creatures.

Of course, the theory of evolution applies directly only to examples of seeming design in biological organisms. Nonetheless, the theory counts against the argument from design in two ways. First, it seems to rob the argument of many of its best examples. Using the theory, for example, one can explain why the various parts of the eye developed in just the way that would best promote good vision, without having to make reference to a designer. Second, even where evolutionary explanations cannot be employed directly, they seem to provide a model for explanations of phenomena that look as if they were designed. For it would seem that purely random processes could result in a universe filled with highly organized and, therefore, one would expect, impressive structures, as long as such structures came about on rare occasions and, once in existence, tended to remain in existence, while less highly organized structures tended to be less stable and more short lived.

I shall attempt to answer these three objections in Part III below. However, for now I shall set them aside and turn to the version of the argument from design I wish to offer.

Part II

The version of the argument from design I shall present here trades heavily on a distinction that is very similar to one worked out by Elizabeth Anscombe in a brief, unpublished paper, entitled 'Cause, chance, and hap',[5] and to a great extent it grew from thoughts stimulated by Miss Anscombe's paper. In 'Cause, chance, and hap' Miss Anscombe distinguishes what she calls 'mere hap' from a sense of 'chance' she defines as 'the unplanned crossing of causes'.[6] When an event occurs by mere hap, there is an element of randomness in its coming about; it might not have occurred, even if all of the conditions relevant to its production had been the same. To use Miss Anscombe's example, the wind might carry a sycamore seed to a certain spot and let it down, though, perhaps, it could have carried the seed just a bit further without anything relevant having been different. And if so, we can say that it merely happened that the seed dropped where it did and not a bit further on. Notice that in this case we will have a violation of the doctrine of determinism, and indeed determinism might be expressed simply as the thesis that nothing ever occurs by mere hap.

There are other sorts of cases, though, where we would say that something happened by chance, though there need be no violation of determinism. To use another example from 'Cause, chance, and hap', a plane might jettison a bomb which hits a boulder as it rolls down a slope. And here Miss Anscombe will say that it was by chance that the bomb hit the boulder, if it was not intended that it should, even though it may be that no randomness was involved. Perhaps, given sufficient information about the path of the boulder and the manner in which the bomb was jettisoned, one could predict with certainty that the bomb was going to hit the boulder or, at least, that it would if nothing intervened.

The distinction on which the argument I shall present below depends is one between Miss Anscombe's sense of 'mere hap' and a notion of chance quite similar to that of an unplanned or unintended crossing of causes. Consider as an example a typical case in which one would say that it was by chance that a friend and I met in a restaurant. One would not mean, in calling this a chance occurrence, that our meeting had no connexion with antecedent causal factors. It might well be that various occurrences brought it about that my friend was at the restaurant when he was and that others brought it about that I was there at the same time. If you wish, it might have been determined, perhaps it even always had been determined, that we would meet. None of this seems to be ruled out when one says that we met by chance. For if it were ruled out, it would be far easier than it is to refute determinism, or, rather, one would not be able to say that cases of this sort occurred by chance, unless one had refuted determinism.

What one does seem to mean in saying that it was by chance that my friend and I met, I think, is that there was no *common* cause of our meeting. I came for whatever reasons I did, and my friend for whatever reasons he did. There was nothing in common to the causal chains that got us there. Thus, if we met at the restaurant because we planned to meet, or if one of us went because he heard the other would be there, our meeting would not be a chance one. And, again, if we both went to the restaurant because a great chef was to give a one-night demonstration or because everyone in our circle of friends was there to celebrate a certain occasion, then, even if we had not intended to meet, one would not be inclined to say that we met by chance. These last two examples show, I think, that it need not be intended that two states of affairs occur together for their co-occurrence not to be by chance. It is enough if the causal chains by which the two come about contain a significant part in common.

In what follows I shall use the term 'chance' in the sense I have described here. More precisely, I shall say that the co-occurrence of two states of affairs comes about by chance when neither is a significant part of the cause of the other, and no third state is a significant part of the cause of both; that is, when the two do not come about by causal chains that have a significant part in common. Also, corresponding to the distinction between mere hap and this sense of 'chance', I shall find it useful to distinguish two ways in which one might explain conjunctions of states of affairs.

Suppose that I happen by chance to be standing up at the moment you read this sentence. Can one then hope to explain why I stand as you read? I think it would be natural to answer in the negative. There is no reason why I was standing as you read; things just happened to work out that way. However, someone who takes seriously the principle of sufficient reason, that every truth has an explanation, will not accept this reply as the last word. Rather, I think, he will want to say that all one has to do to explain the occurrence in question in the sense *he* has in mind is to conjoin explanations of why, at a certain time, I was standing and why, at the same time, you were reading the sentence you were. Corresponding to these two possible answers, let me say that a conjunction of two or more states of affairs, or the co-occurrence of the states of affairs, has a *basic explanation* when and only when each state in the conjunction has an explanation, and I shall say that a basic explanation of a conjunction is a conjunction of explanations of its conjuncts. Further, let me say that the states in a conjunction of two or more states of affairs have an *explanation in common* when and only when explanations of any two states in the conjunction contain a significant part in common. That is, the co-occurrence of a group of states of affairs will be guaranteed a basic explanation whenever no state in the group occurs by mere hap, and the states in the group will have an explanation in common if and only if no two occur together by chance.

Now there are two points I would like to make using these distinctions. First, there would seem to be no logical guarantee that two logically independent states of affairs will have an explanation in common. Things often happen by chance, and a supporter of the principle of sufficient reason can hope for no more than that every conjunction will have a basic explanation. Secondly, however, there are cases in which, as an epistemological matter of fact, we simply would not believe that certain states of affairs had occurred together by chance. And it is on this second point that the argument I shall present is based. For I believe there are natural phenomena which it would be extremely hard to believe occurred together by chance but which, it would seem, could only have an explanation in common if at least some of them were created by design. And thus, I think, by a two-step argument we might be able to prove that some natural phenomena are created intentionally.

I shall not offer an example of the sort of phenomena I have in mind until later. First, let me illustrate the reasoning I hope to use with a fictitious example which, if my memory has not deceived me, is adapted from one I was given a number of years ago by Miss Anscombe.

Suppose that one day a perfect picture, say, of a nativity scene were formed by the frost on someone's window.[7] I think we almost certainly would believe that this occurrence was brought about by design,[8] though not necessarily by the design of a divine being. And if we were asked why, I think we would probably respond that if this were not so, there would be no way to explain why ice formed on the window in the pattern it did. However, if by an explanation, we have in mind a basic explanation, this might well not be true.

Supposedly, in normal cases, various facts about weather conditions, the make-up of a pane of glass, the temperature and humidity in the room in which the pane is installed and the like, cause ice to form in the way it does on a window pane. And also, supposedly, there are possible conditions which, if they were to obtain, might cause ice to form a nativity scene on a given pane. Of course, these conditions might be very strange, but we do not know this. Suppose, in fact, that the nativity scene in our example arose by natural means from conditions that appeared quite normal, that those conditions themselves arose from normal-seeming conditions, etc. Then we can imagine that scientists could give a perfectly good basic explanation of why the pattern formed by the ice in our example was one that constituted a nativity scene.

First, one would explain why ice formed in the pattern it did on the relevant window in the way that one might hope to do so in normal cases, that is, by explaining why ice crystals of various sorts formed on various spots of the window. Then one would explain why the pattern that was formed made up a nativity scene, using facts about geometry, about basic human perceptual mechanisms, perhaps, and the like. The result would be a basic explanation of why the pattern that formed was a nativity scene.

What I think is interesting here, though, is this. If we were given such an explanation of the nativity scene in our example, we would still, I think, be no less inclined than before to believe that it resulted from design. If anything, by showing that the scene arose from processes that were, so to speak, part of the course of nature, such an explanation would make us more inclined to believe that it was designed by a being deserving of our worship and not merely by someone who had made a technological breakthrough over ice.[9]

The question, then, I think, is: what reasoning do we use when we conclude that the nativity scene in our example was produced by design? And the answer, I believe, is the following. First, I think, we believe that ice could not form a nativity scene on a window merely by chance. That is, in our example, there must be an explanation in common of the fact that ice formed the pattern it did on a certain window and the fact that that pattern constitutes a nativity scene. Why we believe this is not completely clear. Ice very often forms beautiful patterns on window panes, and yet we are content to accept that it is by chance that the patterns that are formed are ones that strike us as beautiful. However, that we would not be content to hold similarly that a nativity scene resulted from chance I think is clear.

But if this is true, then the fact that the ice formed a certain pattern on the pane in our example and the fact that that pattern constitutes a nativity scene must share a significant part of their causes in common. And, therefore, either one of the facts is a significant part of the cause of the other, or a third state of affairs is a significant part of the cause of both. However, the fact that a certain pattern forms a nativity scene is a very general one. It results from facts about geometry, about what counts as a nativity scene, and, perhaps, about what patterns we see when we encounter various sorts of objects. And many of these facts are not caused at all, while the remainder, it would seem, as an empirical matter of fact, could be caused neither by the fact that ice formed in a certain pattern on a particular pane of glass nor by the sort of facts, for example, about local weather conditions and the make-up of a pane, that would cause ice to form in such a pattern. And, therefore, it seems that neither the fact that ice formed in a certain pattern nor causes of that fact could be significant parts of what caused the pattern to be a nativity scene.

However, in this case it must be that the fact that a certain pattern constitutes a nativity scene was an important part of what brought it about that the pattern appeared on the window in our example. And this, again as an empirical matter, seems to be something that could not happen unless the pattern was produced by design. For the fact that a pattern forms a nativity scene could give a designer reason to bring it about that it appeared on a window and, thus, play a significant role in an explanation of such a fact. However, if the pattern in our example was not brought about by design, then it seems out of the question to think that the fact that it constitutes a nativity scene might have been an important part of the cause of the very specific conditions holding in and around a particular piece of glass on a particular night that caused it to be formed in ice. And, thus, it seems that the nativity scene in our example must have been produced by design.

The case of the nativity scene, of course, is fictitious, but I believe that similar reasoning might well be able to show that in many actual cases natural phenomena have been produced by design. For in many cases complex states of affairs have come together in ways that have produced noteworthy features of the universe, and one might argue that it could not simply be by chance that they came together in ways that had such impressive results. That is, one might think that there must be an explanation in common of the facts that certain states of affairs have occurred and that, by having done so, they have produced the impressive results they have. However, the fact that various states of affairs would produce impressive results, if they occurred together, cannot, it would seem, be explained by the fact that the states actually did occur, nor by the sort of facts that would cause them to occur. And, therefore, the only alternative is that the fact that the states would produce impressive results helps to explain their occurrence. But, again, it would seem that this could not happen unless the states were caused to occur by a designer acting to produce their impressive results.

This, then, in brief, is the version of the argument from design I wish to put forward, and I want to claim that the reasoning employed in it gains plausibility from the fact that similar reasoning explains our intuitions in the case of the nativity scene. However, I think one might want to dispute this last claim. For, first, it might be thought that our intuitions about the example of the nativity scene are based, not on the complex argument I suggested, but on simple analogical reasoning. That is, one might think that, unlike the sorts of natural phenomena to which arguments from design appeal, the nativity scene in our example is sufficiently like works of art created by human beings for one to reason simply by analogy that it too must have been designed.[10]

And, secondly, one might want to argue that even if our intuitions about the nativity scene can be explained by the reasoning I proposed, that reasoning itself is only a disguised form of analogical reasoning. For it might seem that there is only one way in which one can discover that a group of phenomena have an explanation in common, assuming that one has not discovered this directly by first having discovered the causes of the various members of the group, and that is to find sufficient similarities between the group and others whose members are already known to have explanations in common. But in this case, it might be thought that the conjunctions of phenomena to which an argument from design points differ enough from those whose origins we have been able to discover to block an argument by analogy that their members have explanations in common.

I think, however, that these two objections fail. First of all, it is true that in many respects the nativity scene in our example closely resembles works of art we know to have been designed by human beings. Indeed, the image formed in the example can be supposed to look just like the sort of image artists put on canvas. Nonetheless, there are also ways in which the nativity scene differs from works of art. Most importantly, the process by which the scene came to be on its window differs greatly from that by which paint comes to be on a canvas, and this is just the sort of difference that earlier made us conclude that we could not claim that eyes were produced by design merely because they resemble machines.

That the similarities between our nativity scene and human works of art are not enough to permit a simple argument by analogy for the design of the scene can be shown, I think, by the following consideration. Frequently frost forms beautiful, symmetrical 'snowflake' patterns on window panes, and these patterns often resemble man-made geometrical designs just as closely as the nativity scene in our example might be supposed to resemble works of art. Yet no one would conclude straight away that such frost patterns are produced by design. More than resemblance to works of art, then, is needed to show that the nativity scene in our example would have to have been created intentionally.

Secondly, it might be true that one can infer that phenomena whose causes are not known have an explanation in common, *only* if one does so by analogy from cases that already have been discovered to have such explanations, though shortly I shall suggest that this is not true. But even if it is true, it does not follow that one cannot infer that a group of phenomena have an explanation in common unless there are no significant disanalogies between that group and others that have been discovered to have explanations in common. For it must be remembered that we know phenomena of many different sorts to have explanations in common and, as a result, if a new group of phenomena has important features in common with those groups that have been discovered to have explanations in common, and if the differences between the new group and the other groups seem no more relevant than the differences between some of the other groups themselves, then one should be able to reason by analogy that the members of the new group have an explanation in common. That is, in this case our inference will be more like one which concludes that dogs have a certain feature from the fact that all mammals studied from a wide range of species do, than it will be like one which concludes that dogs have a feature because horses do and because no other mammals have been studied.

At any rate, the fact that the natural phenomena to which arguments from design point are not enough like human artifacts for one to argue by simple analogy that they came about by design is not sufficient to show that one cannot conclude that there must be an explanation in common of the facts that they took place as they did and that, by doing so, they brought about impressive states of affairs. Again, a case in point here is the eye. We cannot argue by simple analogy to machines that our eyes were designed, but it would be preposterous to maintain that the development of the various parts of the eye could be explained without bringing in the fact that they allow one to see. Indeed, one of the most attractive features of the theory of evolution is that it can provide explanations that meet this sort of requirement.

Earlier I suggested that when we infer that two states of affairs have an explanation in common, we might not be reasoning by analogy. In fact, I am inclined to think that the judgement that certain states of affairs have an explanation in common is not dependent on but, rather, is presupposed by inductive reasoning, and I think it might well be the case that it is certain inherent features of states of affairs that make us judge them to have explanations in common, and, further, that we are justified in doing so, if we are justified in reasoning by induction.

Consider a case in which a coin is tossed 100 times and comes up heads each time. One would, no doubt, conclude from this both that the coin had been rigged and that it was almost certain to come up heads on future tosses. And what I think is important here is that these two judgements are connected. In particular, if one thought it was merely by chance that the coin came up the same way on each toss, then it would be as irrational to conclude, by induction, that future tosses would be like previous ones as it would be to conclude, by the gambler's fallacy, that a string of heads must be followed by one of tails. In each new toss the probability would still be one in two that the coin would land heads.

And this, I think, is true in general of inductive inferences. We can sensibly hold that unobserved cases will be like observed ones only when we believe the observed cases have had a significant part of their causes in common. Thus, if we do not believe observed emeralds have been green because of general features of the process by which they were formed, then it would not be reasonable to conclude that unobserved emeralds are green. But if this is true, then it seems that we must be able to make judgements that various phenomena have explanations in common before we can reason by induction and, therefore, it seems that such judgements are not based on induction.

Part III

Let me now turn to the three objections against the argument from design that I mentioned at the end of Part I. Objections (i) and (ii) from Part I tried to show that one could come to know that a state of affairs was brought about by design only directly or by straightforward analogical reasoning. However, if my remarks about our example of the nativity scene were correct, one could know that that scene was produced by design but could not do so in either of these ways. Further, even if I am mistaken, and one could infer by analogy that the scene in our example was designed, still, I think, one could also come to know that it was by the more complex reasoning I described, and such reasoning might let one know that a state of affairs was produced by design, even when this could not be shown by analogy.

Objections (i) and (ii), then, fail, and that they do, I think, should not come as too great a surprise. For the two objections hold, plausibly enough, that one cannot know *a priori* that any given state of affairs was brought about by intentional action. But it does not follow from this that one could know such a thing only directly or by analogy. The objections, then, do not show that there could not be a line of empirical reasoning that is not straightforwardly analogical but that, nonetheless, might allow one to infer that a state of affairs had been brought about by intention. They simply assume that there could be no such line. However, if I am correct, the reasoning I proposed to explain our intuitions about the nativity scene on the window is just such a bit of reasoning, and if the reasoning seems plausible, then objections (i) and (ii) should not stand in its way.

Let me move on, then, to objection (iii). I think many people today are taken by a certain picture of the origin of life in which the theory of evolution plays a large part. As things are represented by this picture, it was simply a matter of good fortune that the earth came to provide an environment suitable for living creatures, though the good fortune here was not particularly surprising. For in a universe as vast as ours there are many stars like the sun, and—often enough—such stars should have planets whose size, composition, and orbit are similar to those of the earth. Then, once the earth afforded the proper environment, the first primitive organisms came into existence as results of what, it is hoped, were not too improbable series of chemical reactions. Again, here, it was simply by chance that the chemical processes that occurred were ones that produced living creatures. Finally, once the first organisms were in existence, it is thought that the theory of evolution can account for the rest and that the mechanisms of chance mutation and natural selection embodied in the theory led to the development of more and more highly developed creatures until, finally, beings evolved that were capable of reason.

This picture of the origin of life seems to be widely held today. Indeed, I believe its popularity is an important feature of the intellectual history of the present age. Nonetheless, I think the picture is flawed. For one thing, we might believe that various chemical processes could produce very simple living creatures, even if the fact that they produced such creatures had nothing to do with the fact that the processes came about. But we would not accept that very complex creatures could come about in this way. However, as Geach has noted,[11] the process of natural selection itself seems to presuppose the existence of creatures with highly developed genetic mechanisms and, so, cannot be used to explain their origin. And, therefore, we must find another plausible account of the origin of these mechanisms.

Natural selection can only take place among creatures that bear offspring that closely resemble their parents without resembling them too closely. For if offspring are exactly like their parents, then natural selection can occur only among characteristics already in existence and, thus, will not lead to the development of new characteristics. On the other hand, if offspring do not closely resemble their parents, then even if certain parents have highly adaptive characteristics and bear many more children than others, their children will not be very likely to inherit the characteristics, and the process will stop.

Of course, in fact creatures do have genetic mechanisms that facilitate natural selection, but the mechanisms are very complicated, and though they might themselves have evolved to some extent by natural selection, it would seem that any mechanism that led to offspring that resembled their parents closely enough but not too closely would have to be very complicated. And so, one would have to ask how they could come about, if not by design. As Geach writes:

> There can be no origin of *species*, as opposed to an Empedoclean chaos of varied monstrosities, unless creatures reproduce pretty much after their kind; the elaborate and ostensibly teleological mechanism of this reproduction logically cannot be explained as a product of evolution by natural selection from among chance variations, for unless the mechanism is presupposed there cannot be any evolution.[12]

Thus, there is much that is noteworthy about the development of living beings that cannot be explained by the theory of evolution. But even if this problem can be surmounted without recourse to a designer, there is a second difficulty.

Simplified accounts of the theory of evolution might make it appear inevitable that creatures evolved with the sorts of impressive and obviously adaptive features that might otherwise be thought to have been designed. For over a sufficient period, one might think, a few individuals would develop such features by chance mutation, and once some creatures had them, the obvious desirability of the features would be enough to explain their proliferation. However, this impression of inevitability, I think, is quite misleading.

Evolutionary change generally proceeds very slowly. We can be confident, for example, that no ancestors of birds suddenly came by wings in a single step and, likewise, that no ancestor of man came to have a brain capable of reason because of one chance mutation. Instead, these sorts of noteworthy and obviously adaptive features come about only as results of long series of evolutionary changes, each of which has to be adaptive and has to become dominant among members of a species, and the noteworthy features themselves cannot come about unless all the others do. Further, these smaller evolutionary changes cannot be counted on to be obviously adaptive, nor always to be adaptive for the same reasons that the larger, more noteworthy changes are. And most importantly, as the term 'adaptive' itself suggests, very often these small changes will be adaptive only because of fine details of, and changes in, the relationship between members of a species and their environment.

Consider the following passage from a recent biology textbook, for example:

> There is . . . good evidence that during the period in which *Australopithecus* lived there existed considerable expanses of lush savannah with scattered shrubs, trees, and grasses. There were berries and roots in abundance, and because such areas were suitable for grazing, these savannahs were well stocked with game. These areas provided new habitats, abundant in food, and so we surmise the australopithecines came down from the trees in which their own apelike ancestors lived in order to avail themselves of these new sources of food . . . . Although descent from the trees does not always result in evolution of upright posture in primates . . . , through a lucky combination of anatomy and habits, these ape-men became bipedal. Being bipedal meant that the hands were freed from locomotor function and could be employed in manipulative skills such as carrying and dragging objects, fashioning tools and weapons, and so on.[13]

This, in turn, led to improvements in the primitive tool-making ability that had preceded upright posture. And finally, 'with the advent of toolmaking, hunting for big game became a possibility, and the brain and the hand were now subject to the molding force of natural selection'.[14]

Now whether the precise details of the picture presented in this passage turn out to be true is not important here. What is important is that something of this sort almost certainly was true. Had not the grass in a certain area grown to the proper height, had not a certain food source become available or unavailable, had not various predators been present or absent, had not climatic conditions been what they were, etc. as ancestors of man developed, human beings would not have come into existence. And if they had not, there seems to be no reason to think other beings capable of reason would have evolved instead. After all, useful as intelligence is, no other species has come into existence with such a high level of it.

Furthermore, seemingly chance occurrences like these did

not play a role only in the final stages of the evolution of human beings. It is likely that, at nearly every step in the evolutionary chain that led from the most primitive of creatures to people, similar sorts of occurrences played a role. In fact, without specific evidence one cannot assume even that it was inevitable that mammals, vertebrates, or even multi-celled creatures would evolve.

But then, one might ask, again, whether it could have been simply by chance that so many seemingly unconnected occurrences came together in just the way that would lead to the evolution of creatures capable of reason, and I think that one might well conclude that it could not have been.[15] At least, it would be very strange, if the myriad occurrences needed to produce human beings came about in just the right way simply by chance and equally strange if the occurrences had an explanation in common, but the fact that they would produce intelligent beings had nothing to do with the fact that they came about. However, one might wonder how so many different sorts of occurrences could have an explanation in common and, indeed, have an explanation in common with the fact that they would lead to the evolution of beings capable of reason, unless they were produced by design?

Of course, I must admit that I cannot prove that the occurrences that led to the development of beings capable of reason could not have taken place by chance. To do so in a fully satisfactory manner, I think, would require a method for distinguishing those conjunctions of states of affairs which require explanations in common from those which do not, and this I do not know how to provide.[16] However, I find it hard to believe that so much could have happened simply by chance, and yet I think this is exactly what one must believe, if one believes that the universe was not created by design. I think, then, that it is safe to conclude that those who fear that the secular view of things, common today among so many intellectuals, robs the world of its mystery are quite mistaken.

Notes

1 See, in particular, the remarks of Philo in the *Dialogues Concerning*

Natural Religion and those put into the mouth of Epicurus in Section XI of *An Inquiry Concerning Human Understanding*.

2 D. Hume, *Dialogues Concerning Natural Religion*, ed. Norman Kemp Smith (Bobbs-Merrill, Indianapolis, 1947), Part II, p. 143.

3 This example was adapted from one suggested to me by David Hills, and I am also indebted to Nancy Cartwright and Paul Humphreys for help in its construction.

4 For Hume's presentation of his thesis see, for example, Section IV of *An Inquiry Concerning Human Understanding*.

5 'Cause, chance and hap' was written by the middle of 1968 and is four pages long in its handwritten manuscript. Much of the material in it, including the notion of 'mere hap', appeared later in Miss Anscombe's 'Causality and determination'. I am relying on a photocopy of the manuscript.

6 This sense of chance and, therefore, the similar one I shall present below seem to be descendants of one discussed by Aristotle. See *Physics*, II, 4–6. I am indebted to Ian Hacking and Paul Humphreys for pointing this out to me and to Terry Penner for the reference.

7 While writing this paper I learned from Miss Anscombe of a similar example which she definitely has used in which a message is spelled out in perfect lettering in the ice on a window. Another, somewhat similar, example is given by Cleanthes in Part III of Hume's *Dialogues*: 'Suppose . . . that an articulate voice were heard in the clouds . . . in the same instant over all nations, and spoke to each nation in its own language and dialect . . . ' (D. Hume, *Dialogues*, *op. cit.* p. 152). Either of these examples and, no doubt, many others besides could be used to make the points I wish to make below with the example of the nativity scene.

8 Anyone who believes a nativity scene could appear on a window without having been designed may substitute a more elaborate example for the one I am presenting. For example, it might be supposed that numerous perfect nativity scenes appear one Christmas morning on the windows of many practising Christians living in cold climates. Remarks similar to those I make below could then be made in connection with this example.

9 It should be noted that an analogous point could be made about the argument from design. Someone holding that certain natural phenomena were designed need not deny that the phenomena resulted from a chain of purely natural causes extending indefinitely far in the past. And, again, if a natural phenomenon one believes to have been produced by design turns out to have resulted from such a chain of causes, that fact may count as evidence for the eminence of its designer.

10 In his introductory material to Hume's *Dialogues*, Norman Kemp Smith makes a similar charge against Cleanthes' use of the example of the voice in the clouds, arguing that the example 'chiefly serves to illustrate Cleanthes' entire failure to recognize the point and force of Philo's criticisms'. (Hume, *Dialogues*, *op. cit.* p. 101).

11 P. T. Geach, 'An irrelevance of omnipotence', *Philosophy* vol. 48, No. 186 (1973), 327–33.

12 P. T. Geach, *ibid.* p. 330.

13 I.W. Sherman and V. G. Sherman, *Biology: A Human Approach* (OUP, New York, 1975), p. 456.

14 I. W. Sherman and V. G. Sherman, *ibid.* p. 456.

15 Of course, one might conclude that such occurrences could have come together by chance. In particular one might argue that, unlikely as it may be that all the conditions needed to produce beings capable of reason should have arisen on earth, still in a universe as vast as ours we can expect that it should have happened somewhere, and earth just happens to be a place where it did happen. However, I think this argument is little more than an appeal to scepticism. One could as well argue that we do not know that the speed of light is constant in a vacuum, because if the speed of light were random one still would expect it to appear constant in some region or other of a large enough universe. The point in both cases is the same. Random processes can be imagined and—in a large enough universe—expected to mimic controlled processes, but when phenomena of the right sort would be sufficiently improbable if they occurred by chance, we have a right to conclude that they did not occur by chance.

16 The problem here, I think, is closely connected with *one* of the problems of induction. If a coin is tossed 1,000 times, and the results form certain patterns of heads and tails (for example, if the coin lands heads on all and only the prime numbered tosses), then we will believe that the pattern that was formed did not occur by chance, and we will expect future tosses to result in a similar pattern. On the other hand, other patterns of heads and tails would strike us as 'random', and one would not expect them to be repeated by future tosses. But how do we distinguish the random patterns from the others?

[9] INTENTIONS IN THE CONDUCT OF THE JUST WAR

Lucy Brown

In the course of her life Professor Anscombe has published three essays relating war and murder, *The Justice of the Present War Examined* (1939), written in collaboration with Norman Daniel,[1] *Mr Truman's Degree* (1956), and 'War and murder' (1961).[2] It is proposed in this essay to put the development of her ideas into their context, and then to consider their practical application in historical situations.

The first essay was written in the opening weeks of the last war. It is datable by the fact that Poland is mentioned as still fighting: in fact the Polish government retreated to Rumania on 19 September, 1939. It was written, in a smooth and grave manner quite unlike her later writings, while she was an undergraduate, and must be her first publication. It presents the traditional conditions, derived from the natural moral law 'which must all be fulfilled for a war to be just':

> (i) There must be a just occasion: that is, there must be a violation of, or attack upon, strict rights.
> (ii) The war must be made by a lawful authority: that is, when there is no higher authority, a sovereign state.
> (iii) The warring state must have an upright intention in making war: it must not declare war in order to obtain, or inflict, anything unjust.
> (iv) Only right means must be used in the conduct of the war.
> (v) War must be the only possible means of righting the wrong done.
> (vi) There must be a reasonable hope of victory.
> (vii) The probable good must outweigh the probable evil effects of the war.

The pamphlet proceeds to consider whether the war fulfils those conditions. The occasion, she says, is just, since the rights of Poland have been infringed, and it is being waged by a lawful authority. 'There is, so far as we can tell, a reasonable hope of victory'—an astonishing view at a time when the Nazi–Soviet pact was in being, particularly if the specific aim was the restoration of Poland's frontiers and independence. She accepted that the wrong might not be righted by other means. The remainder of the essay discussed her doubts on conditions (iii), (iv) and (vii). Was the championship of Poland sincere, or was it a pretext for 'a policy, not of opposing German injustice, but of trying to preserve the status quo and that an unjust one'? Here she repeated, in an extreme form, the anti-Versailles arguments which were such a powerful influence on public opinion between the wars, that is, that Germany had been unjustly treated in 1919 and was now taking by force what should have been agreed by negotiation. This might have been fair enough up to, say, 1936, but she seems not to have understood that in Austria and Czechoslovakia Germany had already gone far beyond rectifying anything in the Treaty of Versailles. Nor, which is also strange, did she anywhere suggest that Hitler was dangerous, either to the Jews, or to the nations round about. Being at that time (I imagine) uninterested in politics, she had absorbed the liberal doctrine that war propaganda distorts, and gave Hitler the kind of benefit of the doubt that might have been more appropriate to the Kaiser.

Professor Anscombe's deductions from previous events about the government's intentions were poorly based, but her assessment from what they were currently saying about their aims was acute and prescient. 'For the truth is', she wrote,[3] in September, 1939, when the war was only a week or two old:

> that the government's professed intentions are not merely vague, but unlimited. They have not said: "When justice is done on points A, B and C, then we will stop fighting." They have talked about "sweeping away everything that Hitlerism stands for" and about "building a new order in Europe". What does this mean but that our intentions are so unlimited that there is no point at which we or the Germans could say to our government: "Stop fighting; for your conditions are satisfied".

This passage deserves attention. She has repeated it in various forms ever since, and criticism of the demand for unconditional surrender is now common. But it was very unorthodox then: this description, also of undergraduate opinion in Oxford in September, 1939, illustrates a more common attitude:

> No one can study the propaganda of this war without noticing its wide international note. Broad general ends are spoken of, freedoms, ways of life and so on. What we are fighting against is carefully described; what we are fighting for is expressed in more general terms, the interests of one

nation alone, the honour of one nation alone, does not seem to be sufficient.[4]

The Justice of the Present War Examined proceeded to discuss the major question of this and of her subsequent writings on the subject: what are and are not permissible ways of waging war. The argument may perhaps be summarised as follows:

(1) 'It is not lawful to kill men simply punitively, except after trial'.[5]

(2) 'It is no sin to kill a man in self-defence or in defence of rights, if there is no possibility of appeal to a higher authority'.[6]

(3) 'The actions of a great mass of the civilian population are not in themselves wrongful attacks on us. Therefore they may not be killed by us, simply as deserving to die, nor yet because their death would be useful to us'.[7]

The idea that war is indivisible was attacked both here[8] and in her later writings:

> Civilians are not committing wrong acts against those who are defending or restoring rights. They are maintaining the economic and social strength of a nation, and that is not wrong, even though that strength is being used by their government as the essential backing of an army unjustly fighting in the field.

There was, she believed, sufficient evidence already that these principles would be disregarded. Blockade is indiscriminate in its effects. The bombing of civilians was probable. At the outbreak of the war Roosevelt had asked for guarantees that civilians would not be attacked: the British agreed, but added the proviso that they would 'adopt appropriate measures' if the Germans resorted to the bombing of civilians.

The essay concluded with a general discussion of the probable balance of good and evil. No good consequences were foreseen; the evil consequences would be the repetition of the cycle of the harsh vindictive peace leading to future German resentment and future wars.

This is a bald summary of the argument which unhesitatingly denounced the war. It was not based on any military or political assessment of the situation—otherwise Professor Anscombe could not have thought that a war for the righting of Poland's wrongs had a good prospect of victory. An armchair strategist might also well have thought that the reason for expecting indiscriminate bombing was that it might be the only form of quick intervention on behalf of Poland that was possible at that time. Yet by reasoning from the seven principles which had been listed the pamphlet diagnosed and attacked what were to become two of the most disputed features of the second world war, the bombing offensive on Germany and the demand for unconditional surrender. It is a characteristic and impressive essay.

The second essay dates from 1956. In the summer ex-President Truman was to visit England, and it was proposed to give him an honorary degree at Oxford. Professor Anscombe decided to oppose the decree in Convocation, on the ground that he had signed the order to drop the atomic bombs on Hiroshima and Nagasaki, at a time when the Japanese were suing for peace—the extreme of indiscriminate warfare. She spoke against it on 2 May, 1956 but the decree was carried. *Mr Truman's Degree* was published in the following autumn to explain her reasons.

The most remarkable feature of her action was its failure to attract support. To write against the war in 1939 may have been an eccentric thing to do, but in 1956 Professor Anscombe was swimming with the tide: there was a great deal of agitation against nuclear testing and against possession of the bomb, though the reasons for opposition might vary. (The first of the Aldermaston Easter marches took place in 1958.) Nor were people in general unduly timid about causing offence to other countries: in April, 1956 Bulganin and Khrushchev, effective and not ex-rulers, had visited Britain and there had been a great deal of criticism and public protest, people complaining that they were being asked to 'shake hands with murder'. No doubt many members of Convocation had given no particular thought to Mr Truman's claims, but it might have been expected that, the grounds of opposition once stated, there would have been a handful of Quakers, fellow-travellers, experts in radiation hazards and so on who would have felt in honour bound to support her, though from different points of view.[9] However they did not. Her action was reported briefly in *The Times* but attracted none of the interest which had once been given to the King and Country debate of 1933, though it is quite as significant, negatively, as an indication of public opinion. The episode is also a reminder of how greatly universities have changed since then: had it happened a dozen years later, Professor Anscombe would have been accompanied by a crowd chanting slogans with which she might not agree.

Mr Truman's Degree expounded the distinction between killing and murder in the same way as *The Justice of the Present War Examined* had done. In other ways there was a change of emphasis. The argument was much more firmly tied to the history of the war and the dropping of the bomb, as was necessary if she was to persuade a general audience who had memories of the events. And, even though the development of area bombing had fulfilled the direst prophecies of 1939, she was now clearly more sympathetic to the purposes of the war than she had been before. She still argued that the demand for unconditional surrender pushed the fighting to more frightful extremes, but she added 'Now the demand for unconditional surrender was mixed up with a determination to make no peace with Hitler's government. In view of the character of Hitler's regime that attitude was very intelligible.'[10] She could no longer be faulted, as she might have been in 1939, on the grounds that she did not understand what the enemy was like.

In two other ways the argument had changed in emphasis. It was centred on the distinction between killing and murder, and rejected, just as forcibly, the notion that war is indivisible. But the handling of the traditional principle of double effect was a little different. In 1939 Professor Anscombe described it as 'exemplified when an action designed to produce one effect produces another as well by accident'.[11] She had then proceeded to limit the use of the principle, 'No action can be excused whose consequences involve a greater evil than the good of the action itself, whether these consequences are accidental or not'. Also, 'It is a different thing, while making one group of persons a target, to kill others by accident, and to make a group of persons a target, in order—by attacking them all—to attack some members of the group who are persons who may legitimately be attacked.' The loopholes were being closed. In *Mr Truman's Degree* the principle is expounded with its important rider:

> 'For killing the innocent, even if you know as a matter of statistical certainty that the things you do involve it, is not necessarily murder. I mean that, if you attack a lot of military targets, such as munitions factories and naval dockyards, as carefully as you can, you will be certain to kill a number of innocent people, but that is not murder. On the other hand, unscrupulousness in considering the possibilities turns it into murder.[12]

The difference is one of tone, rather than of doctrine.

Secondly, and giving the clearest indication of a shift in position, *Mr Truman's Degree* includes a clear-cut attack on pacifism as a false and dangerous doctrine; false because it denied the right of the state to fight its external enemies in just circumstances, and dangerous because, she said, by saying that all war was bad it blurred the distinction between killing and murder. In this attack there was nothing incompatible with the arguments of 1939, which did not include any objections to war in general but merely to the war then beginning.

The third essay, 'War and murder', was published in 1961 in a collection *Nuclear Weapons: A Catholic Response.* These circumstances may have helped to dictate its form: Professor Anscombe was clearly writing with Catholic readers in mind. She was also apparently concerned to separate herself from enemies on either flank, from cold-war hawks who would not be sorry to see a just war against Communism and from Christian pacifists who held 'a conception of Christianity as having revealed that the defeat of evil must always be by pure love without coercion'.[13] Far more than either of the previous essays, 'War and murder' was concerned to defend the right of the state to use violence: 'For society is essential to human good; and society without coercive power is generally impossible.' This had, presumably, hardly needed saying on the previous occasions. The state's right to oppose its external as well as its internal enemies was again argued, but here again there was a slight, scarcely perceptible, strengthening of the claim. In 1956 she had written:

> The right to order to fight for the sake of other people's wrongs, to put

right something affecting people who are not actually under the protection of the state, is a rather more dubious thing obviously, but it exists because of the common sympathy of human beings whereby one feels for one's neighbour if he is attacked. So in an attenuated sense it can be said that something that belongs to, or concerns, one is attacked if anybody is unjustly attacked or maltreated.[14]

There was nothing 'dubious' or 'attenuated' about the claim in 1961: 'Further, there being such a thing as the common good of mankind, and visible criminality against it, how can we doubt the excellence of [the British suppression of the slave trade]?'

The essay continued with a further attack on pacifism and with a discussion of the principle of double effect. Both sections were directed to Catholic readers and do not mark any change from what had been written in *Mr Truman's Degree*. (The first deals with reasons why pacifism is no part of the Christian tradition; the second with the misuse of the doctrine of double effect.)

Professor Anscombe has travelled some way: her central principles have not changed and could hardly do so. She is like a painter who repeatedly returns to the same subject, emphasizing and defining some of its features and losing interest in others. The practical lessons have changed a great deal. In 1939, faced with the fact of war, she suspected that it would lead to murder and was against it: now she writes in favour of wars for just purposes—which need not be solely defensive. In so doing she sometimes writes as if legitimate war had a manageability and acceptability which is not often in evidence.

Pacifists have become a main target. They are blamed, not just for their doctrines but for their influence on opinion: 'How endlessly pacifists argue that all war must be *à outrance*! that those who wage war must go as far as technological advance permits in the destruction of the enemy's people.'[15] Pacifists are not an organized body with a defined set of beliefs and we cannot pin them down firmly. But if they mean 'must' in the sense that wars do go as far as technological advance permits they have a lot of history on their side: the crossbow, gunpowder, and the Maxim gun were greeted with a shudder when they first appeared, but were put to use none the less, not necessarily against the enemy's people. In war, as Professor

Anscombe herself shows, there is no common authority to adjudicate between the parties, and the penalties of failure are very great. This makes advance to the technological limit the more probable.

I would suggest that the breakdown of the distinction between killing and murder in war (which has certainly occurred) has a different origin. There was a great deal of anti-war feeling in the 1930s, but it was not all an expression of a coherent pacifism (if it had been so this country would never have accepted conscription in April, 1939, before the outbreak of war, with scarcely a protest). When in the 1930s people said 'Never again', they were saying it to the trench warfare of the western front. In the static warfare of 1914–18 the combatants were physically separated, more clearly than in many wars, from the non-combatants: such fighting fell within the rules—killing not murder, but on an unprecedented scale. The pacifist, at last goaded to defend himself, might reply to Professor Anscombe that she is so concerned with blackening murder that killing is scarcely mentioned as objectionable. But in the countries of western Europe the memories of 1914–18 were such that every country was determined to find some alternative to trench warfare in future, whether by the development of tanks or aircraft. This provides the background to two otherwise ill-assorted features of the 1930s: that foreign policy was weak and appeasing until far too late, and that rearmament, when at last it came, put its cash and faith in an independent bomber command, which could so easily be used indiscriminately. During the war 'Bomber' Harris used the argument, when asking for more bombers, that they were the alternative to 'the mud of Flanders and France'.[16] It does not follow, of course, that these were the only alternatives, or that area bombing was the only use to which bombers could be put. It is legitimate to argue that, had there been a more alert and critical public opinion (one less corrupted by pacifism), the slide from selective to area bombing might possibly not have taken place. In all this pacifism appears to play a minor role, the major one being taken by the scale and destructiveness of legitimate warfare between 1914 and 1918.

Doubts must persist about how the principle of double effect could work in practice in war. It is held[17] to excuse incidental killing on strict conditions:

(1) The act itself must be morally good or at least indifferent.
(2) The agent may not positively will the bad effect.
(3) The good effect must flow from the action at least as immediately as the bad effect.
(4) The good effect must be sufficiently desirable to compensate for the bad effect.

In her essays of 1956 and 1961 Professor Anscombe replied to critics that there will be borderline cases in deciding whether particular acts of war are legitimate or not: the trouble is that so many of the practical situations of war would fall on this borderline, and it is a borderline, not between different grades of more or less acceptable behaviour, but between what is permissible, possibly just, and murder. For example, in 1939 Professor Anscombe cited as an improper use of the principle of double effect shooting at a group in the hope of hitting the offenders.[18] It seems to me that again and again in wars, both present and past, situations arise where a person of scrupulous conscience would be unable to decide whether or not this was what he was about to do. The siege of a city is subject to the same objections as a blockade, with the difference that the city, if it has a garrison, offers a clear military threat to an advancing army. It is argued that it is legitimate to bomb munition factories and supply lines.[19] The latter do not discriminate between users: the attacks in World War II on communications and oil supplies in Germany have been shown to have been more damaging to civilian life than direct attack on cities.[20] Munitions production is scattered all over the place in large and small factories: I recently read of one on an island in the Thames at Richmond. To attack so small a target in a dormitory suburb might not seem legitimate, yet the factory might be, or be thought to be, producing some vital component. How would the scrupulous attacker interpret the fourth of the conditions? Civilians and military targets are geographically entwined with each other, and the organization of production and distribution today is such that a spanner in the works can have consequences in far removed areas of life: this much has been made clear by the strikes of recent years. It can be hard to construct plans of campaign that will not injure civilians at some stage.

Non-combatants suffer from war in ways which are far more

various than intentional murder or incidental killing. Until comparatively recent times soldiers lived off the country. Complaints in the past were as likely, or more likely, to be of the depredations of the troops as of their attacks on human beings. An army can commandeer transport, it can pursue an enemy through growing crops, it can buy its food from the local population, or it can steal it, or in an orgy of hooliganism[21] lay waste the fields and drive off the animals. Or, because they have come from a distance, they can spread epidemics among a population which has developed no resistance to them. Some of these acts may be honourable in intention, some parts of the accepted conduct of war, some wicked but not murderous, and some involuntary and uncontrollable. Their combined effects may be slight, or they may be catastrophic: it will depend on the weather or the harvest or whether the troops reappear next year. But in places where transport is primitive and the people are living near the margin of subsistence (that is, in most of the world for most of history) the mere presence of troops and their supply problems constitute a danger to the innocent. If resources are diverted to war they are diverted from something else. In the expedition to rescue Gordon from Khartoum in 1884 for example, General Wolseley's main preoccupation was to buy every possible camel at inflated prices from obliging sheiks. They were absolutely necessary to the success of the expedition: we do not know how the people got on whose normal supply system was thus removed. Another well-known example comes from World War II. In order to supply the North African campaign, shipping was diverted from the Indian Ocean to the Mediterranean. An Indian famine followed which was estimated to have killed 1.5 million people, five times the casualties caused by the bombing of Germany.

To recite these things is not to be a pacifist, nor to condone murder: it merely remembers that they happen. In war there seems to be a wide gap between intentions and effects, so that respectable intentions can lead to atrocious results; misleading information, unprecedented situations, lack of training and inaccurate equipment all playing a part. This is a reason why the doctrine of double effect is difficult to apply. The textbook illustrations in which it is illustrated are limited and clear-cut. The doctor administering a pain-killing drug which may kill the patient has a relatively straightforward problem: it is assumed that he knows the properties of the drug and the condition it is prescribed for and that it will be administered to the right person in the right way. Military calculations are subject to far wider uncertainties; the rarity with which plans and forecasts prove reliable is some evidence of that. On the other hand various usually bad effects are, not invariably, but very usually, consequent on a state of war: invasion involves the disruption of the normal processes of government; plough-shares are abandoned for swords; the civilian population is put at risk. The list is not exhaustive. How often can such consequences honestly be covered by the principle of double effect, and how often is it that 'unscrupulousness in considering the possibilities turns it into murder'?[22] A ruler contemplating war knows very well that these risks exist.

In her later writings Professor Anscombe is concerned to defend the idea of a just war, but she does not sufficiently emphasize, I think, how rarely a just war can happen in practice. There are wars which pass all tests: she quotes the example of the suppression of the slave trade as one, and the Entebbe raid might be another. In these two examples the greatly superior skill and resources on the side of justice would help to reduce the uncertainties of the operation, and thus keep incidental effects within bounds. But such examples are rare.

It might be replied that this point savours of Professor Anscombe's 'hypocrisy of the ideal standard', and that by observance of the rules most wars could be conducted less murderously. But the rules which separate killing and murder are not clear in their operation, and they have to be applied in daily decisions by average people. Thus a war, once embarked on, is likely to lead to injustice.

On the other hand, the seven conditions she cited in 1939—but not as a group subsequently—are much clearer in their combined implications. It is true that they too present difficulties in their specific application. The answers to condition (vi) that there must be a reasonable hope of victory, for example, would vary greatly—even if given in good faith—according to the information and temperament of the analyst. Ordinary citizens rarely possess the political knowledge to carry out the exercise effectively. Furthermore, there may be no true answer to some of the questions. There is usually an answer to the question 'Who is the aggressor?', but not so often to the question, 'Who is in the right?'[23] For example country A cannot control its semi-independent tribesmen who harry the villages of country B, while country B has no power to reduce the vexatious restrictions on foreign traders imposed by its southern seaports. Both have just grievances. The border fighting between England and Scotland in the middle ages, or the War of Jenkins' Ear were of this kind. Here again the political relations between countries are far more complicated than relations between individuals. Nevertheless, in spite of such objections, the general tenor of the rules is clear to any ruler who tries to conduct his affairs justly: war is best avoided altogether; particularly in condition (v) where war must be the only possible means of righting the wrong done. This lesson does not emerge so clearly from Professor Anscombe's later writings.

The treatment of the just war is like much political theory in that discussion moves from the relations of individuals—'Suppose I say to you . . . .'—via larger but familiar associations (for example the club or the college) to the whole scene of international relations, past and present. At each stage, in real life, there are different circumstances to consider (for example the country where the legitimate government is no longer in effective control). A body of international rules which is grounded in the individual's right to defend himself and lack of right to punish his attacker in the process, has a clear general message, but could not be worked to in a precise way. In this respect the conditions of a just war are unlike a body of law which is constantly exposed to actual difficulties and exceptions in the daily practice of the courts.

Notes

1 It is a pamphlet in two separate sections of which the first is by Professor Anscombe.

2 Published in *Nuclear Weapons: A Catholic Response* ed. Walter Stein (Merlin Press London, 1961) and reprinted in *War and Morality* ed. R. A. Wasserstrom (Wadsworth, California, 1970). References in this present essay are to the 1970 reprint.

3 G. E. M. Anscombe, *The Justice of the Present War Examined*, p. 10.
4 R. B. McCallum, *Public Opinion and the Last Peace* (Oxford, 1944), p. 179.
5 G. E. M. Anscombe, *The Justice of the Present War Examined*, p. 15.
6 G. E. M. Anscombe, *ibid.* p. 14.
7 G. E. M. Anscombe, *ibid.* p. 17.
8 G. E. M. Anscombe, *ibid.* p. 16.
9 Depending on what she actually *said.* It is hard to imagine a speech stating what had once been principles of international law which would antagonize these groups.
10 G. E. M. Anscombe, *Mr Truman's Degree* (Oxford, 1956), p. 1.
11 G. E. M. Anscombe, *The Justice of the Present War Examined*, p. 17.
12 G. E. M. Anscombe, *Mr Truman's Degree*, p. 6.
13 G. E. M. Anscombe, 'War and murder', pp. 52–53.
14 G. E. M. Anscombe, *Mr Truman's Degree*, p. 10.
15 G. E. M. Anscombe, 'War and murder', p. 49.
16 Webster and Frankland, *The Strategic Offensive Against Germany, 1939–1945* (vol. I, 1961); Harris to Churchill, 16 June 1942, p. 341.
17 These are taken from McGraw-Hill's *New Catholic Encyclopedia*, entry on 'Double effect'.
18 G. E. M. Anscombe, *The Justice of the Present War Examined*, p. 18.
19 G. E. M. Anscombe, 'War and murder', p. 45.
20 Webster and Frankland, *op. cit.* (vol. III, 1967), p. 244ff.
21 In Western Europe in the Middle Ages this was the normal way of waging war. I am grateful to Dr D. P. Waley for information on this and a number of other points.
22 See p. 138 of this volume.
23 See G. E. M. Anscombe, 'War and murder', p. 44.

[10] INTENTION AND SEX
Jenny Teichman

Professor Anscombe has written three papers on the subject of contraception. In them she sets up certain concepts of *act* and *intentional action* with which she constructs a powerful, albeit somewhat intricate, defence of traditional Catholic teaching on this question.

The first paper, called 'You can have sex without children: Christianity and the new offer', was presented to the Canadian Centenary Theological Congress in 1967 and subsequently published in the proceedings of the Congress.[1] The second paper, 'Contraception and chastity', was read to a meeting of the Bristol Newman Circle and in 1972 published in the journal *The Human World*.[2] The *Human World* paper provoked two replies: a short letter from Peter Winch and a longer letter signed by Bernard Williams and Michael Tanner. Both letters appeared in the journal later in 1972 together with an answer from Professor Anscombe.[3] In 1977 a slightly shortened and slightly altered version of 'Contraception and chastity' was published as a Catholic Truth Society booklet.[4] Finally, in 1978 Professor Anscombe read a paper called 'On Humanae Vitae' to a congress of Catholic doctors and nurses in Melbourne, Australia.[5] In the present article I will refer to these papers as follows: as 'CC' (the first paper—Canadian Centenary), 'HW' (*Human World*—the second paper in its first version), 'CTS' (Catholic Truth Society—the second paper in its second version), and 'OHV' (On *Humanae Vitae*—the third paper).

Professor Anscombe begins CC by saying that the traditional Catholic stand against programmes for large-scale teaching of contraceptive methods should be given up 'if it is indeed true that abortion in our time is pandemic'. For 'it may be desirable to have contraceptive programmes as we have brothels—to avoid worse evils . . .' Nevertheless, a lesser evil is still an evil and must be recognized as such so that those who wish to act rightly can learn what to avoid.

Why then has contraception always been regarded as an evil by Catholic theologians; what are the reasons behind this teaching?

In CC Professor Anscombe writes: 'The ground for counting such an act wrong and shameful was formerly that it was not an act of natural intercourse . . . one might define as "sins against nature" complete sexual acts which deviate from complete acts of ordinary intercourse.' However, as she points out, this ground applies to some varieties of contraceptive practice but not to all. Thus a woman who has taken a contraceptive pill can go ahead with a complete act of intercourse, one that is not deviant, without risking pregnancy.

In HW and CTS another suggestion is made. In the Dark Ages when the biology of reproduction was not understood it was believed that the male seed contained the whole child: no one knew about the existence of the ovum. On this view destruction of the seed would be homicide, or at the least, abortion. The Dark Ages objection still holds, of course, against those types of pill or device which do in fact induce abortion but it cannot stand against non-abortificient types of contraception.

The general objection to mechanical devices is that they render the physical act deviant. The objections to pills or devices which induce abortion are the same as those to abortion itself (and these objections are not the subject of the present discussion). But what objections can there be to bathing and douching (say), or to non-abortificient pills?

One possible objection, namely that pills are themselves artificial things, is rejected by Professor Anscombe. In her HW answer to Williams and Tanner she writes: 'Strange supposition, that Catholic thinkers would distinguish between a pharmaceutical invention and a herb included in your meals for contraceptive purposes!'

Another objection—that it would be just plain ridiculous to forbid some contraceptives and allow others—is taken seriously. In HW she writes: 'This [i.e., drawing lines between methods] would have been absurd teaching; nor have the innovators ever proposed it.'

Now, the assertion that drawing distinctions is (in a given case) an absurd thing to do, cannot stand on its own feet (whatever the given case)—it needs the support of an argument designed to show that differences in method or means make no difference to the nature of the act. For example, it is sometimes said that abortion and contraception are really the same sort of thing, on the ground that when one is allowed sooner or later the other will be allowed too.[6] This is like saying that all war is unjust, on the ground that an initial resolve to use only just methods of warfare rarely prevents a nation from turning to unjust methods when victory looks to be in the balance. Pacifists who point out how nations go on *in practice* ought not to be accused of failing to make *theoretical* distinctions; and conversely, it won't do to reject attempts to draw distinctions between contraceptive methods merely on the supposition that people, or governments, may find such distinctions difficult to understand, or merely on the ground that drawing distinctions is liable to provoke ridicule. As I have already said, the drawing of a distinction between methods can only be rejected if it is shown that the distinction makes no difference to the essential nature of the act. An argument to show just this is in fact provided by Professor Anscombe.

In CC she divides her central question into two questions:

(i) What characteristic of artificially contraceptive sexual acts is condemnable?

(ii) What makes this characteristic a condemnable characteristic?

To begin with some wrong answers to (i) are cleared out of the way. It is not that sex is bad in itself: 'the austere and grudging attitude of older authors towards sex has already been abandoned in the modern Church'. The correct teaching must be St Augustine's when he writes that sex cannot be a bad thing since it is the source of human society.

It is not that the sexual act ought only to be performed when there is a positive intention to get a child. Within marriage sex when procreation is impossible is not forbidden by the Church. Professor Anscombe does indeed seem a bit uneasy about the fact that St Augustine also writes that for married couples to indulge in sex 'purely for pleasure' is sinful. She concludes, though, that when the Church has condemned 'sex for pleasure' what it meant to condemn was sexual behaviour which is against wisdom, or immoderate, or seriously inconsiderate of the other partner, or a manifestation of a preoccupation with sex, or a manifestation of lack of self-control. In HW she says that when the act is governed by a reasonable mind one 'may rightly and reasonably be willing to respond to the promptings of desire' and that desire 'is for intercourse as pleasurable'.

It is not that the intention to avoid conception is invariably a condemnable intention. The Church allows that conception may be avoided by married couples 'for grave reasons', either by abstinence or by the use of safe period sex. Professor Anscombe says that a marriage which is childless, not because of accidental infertility, but because the marriage partners have deliberately made it so (by any means) is no marriage at all: but as far as I can see she cannot and does not object to *family planning*—as long as there *is* a family, so to speak. 'Spacing' children in order (say) to preserve the health of a frail woman; or to make possible the carrying-out of some important work commitment of the husband or of the wife; such spacing, it seems, is not condemnable according to Catholic teaching as expounded here.

What then is the characteristic of artificial contraceptive acts that makes them condemnable? In order to answer this question a sort of schema is set up which could be represented in a diagram like Fig. 10.1 which follows. Although Professor Anscombe does not represent her thought diagrammatically I do not think that Fig. 10.1 distorts her line of reasoning.

The explanation of the schema is as follows: human sexual acts can be classified as either normal or deviant; normal sex acts regarded just as physical acts are *intrinsically generative*; deviant sexual acts regarded just as physical acts are *intrinsically non-generative*. To say that a normal sex act is intrinsically generative does not mean that all such acts must result in generation. In fact many, perhaps most, do not, just as many, perhaps most, of the acorns in a forest will fail to grow into oak trees. But for all that an acorn is the *type of thing* that is generative (of oaks); and a normal human sex act is the *type of act* which is generative (of human beings). An abnormal deviant or non-natural sex act is the *type of act* which, though sexual in character, is intrinsically inapt for generation, perhaps rather in the way that oak settees and seedcakes,

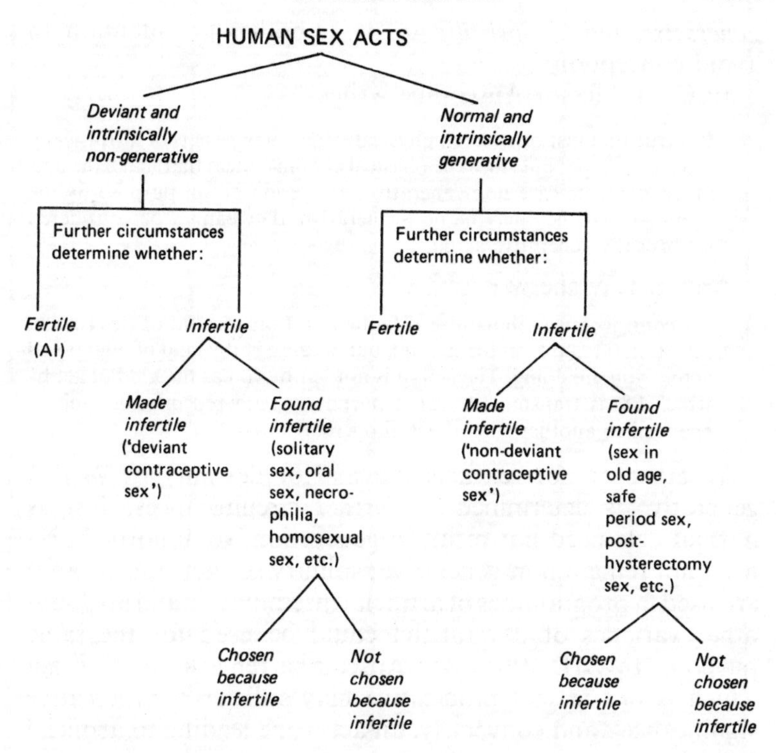

Fig. 10.1

though oakish and seedish as the case may be, are yet intrinsically non-generative kinds of things.

Now, an act of 'contraceptive sex' (to borrow Professor Anscombe's convenient shorthand) may or may not be intrinsically generative *qua* physical act. It is not intrinsically generative *qua* physical act if it involves the use of mechanical or chemical barriers; for the presence of such barriers alters the 'shape' of the physical act and makes it into a different type of physical happening from a normal act. The act becomes one of deviant sex. On the other hand, an act of contraceptive sex which utilizes (say) a pill which is taken beforehand, or a douche which is administered afterwards, is not affected in its 'shape' or character *qua* physical act by these things which happen before or after the act itself. It is not deviant *qua* physical act, *but regarded as an intentional act it is non-*

generative and intrinsically so. It embodies an intention to avoid conception.[7]

In CC Professor Anscombe writes:

> It is true that just considered physically they may be acts of intrinsically generative type; but since the physical circumstances that make the acts in the concrete case non-generative, are produced on purpose by the agent so that they may be non-generative, they cannot be considered intrinsically generative as *intentional actions*.

And in HW she writes:

> . . . contraceptive intercourse is faulted not on account of this further intention (i.e., to limit the family), but because of the kind of intentional action you are doing. The action is not left by you as the kind of act by which life is transmitted, but is purposely rendered infertile, and so changed to another sort of act altogether.

Whether or not sex acts (normal or deviant) are *in fact* generative is determined by further circumstances. Just as normal acts need not result in generation, so abnormal acts need not remain non-generative; solitary sex acts for instance are used in programmes of artificial insemination and no doubt other varieties of deviant act could be used for the same purpose. In OHV Professor Anscombe remarks that an act which is not in fact procreative may still have 'procreative significance', and conversely, an act—one leading to artificial insemination, say—may be procreative yet lack 'procreative significance'. Artificial insemination, she says, is not the same as begetting: 'Begetting is a personal act involving the actual union of man and woman'. (Perhaps the following is an analogy: intravenous feeding is *nutritive*, but it is not exactly the same thing as *eating*.)

Infertile deviant acts are of two main kinds: those made infertile by human action—i.e., contraceptive acts involving devices which alter the physical shape or pattern of the act—and those infertile to start with, the 'acts intrinsically inapt for generation'. Professor Anscombe is emphatic that deviant sex is condemnable because, and only because, it is intrinsically non-generative. She is thus able to state, in CTS: '. . . if contraceptive intercourse is all right then so are all forms of sexual activity'.

This cannot possibly be right, since we know that there are perversions which necessarily involve cruelty and even murder—for instance, child-rape, necrophilia. These things are not wrong because (or merely because) they are intrinsically non-generative. Again, there are at least three different varieties of deviance in sexual behaviour:

(i) deviance consisting in the fact that wrong bits of the body are used (e.g., sodomy);
(ii) deviance consisting in the fact that you have the wrong kind of partner (e.g., bestiality); and
(iii) deviance consisting in the fact that an otherwise normal act is preceded by or surrounded by cruel or bizarre accompaniments.

It is not inconceivable that these differences might form the basis of some important moral distinction or other. Finally, there is a very widespread feeling that some kinds of deviant sex are more deviant, worse, than others. The view that some or all acts usually called deviant are not evil at all, and the view that all such acts are equally evil, are both of them less common than the view that there are degrees of deviance. Indeed, Professor Anscombe, when she uses the expression '*very* deviant' appears to support this latter opinion. But if the evil lies only in the fact that deviant acts are intrinsically non-generative then it cannot be a matter of degree.

Some normal sex acts result in conception and some do not. Normal sex acts which are known to be infertile are either acts which have been made infertile by means which have not altered the physical shape of the act—what we might call 'non-deviant contraceptive sex acts'—or are acts, like old age sex, which are infertile for natural physiological reasons. This last group can be subdivided into those acts chosen just because they are infertile and those chosen for some other reason. Family limitation by use of safe period sex is an example of choosing a physically normal but naturally infertile act because it is known to be infertile.

The characteristic which makes all contraceptive sex acts condemnable therefore turns out to be as follows: in contraception the sexual act has been altered on purpose—by something done at the time, or before, or afterwards—in such a way as to make it into a different kind of action, one that *embodies the intention* of avoiding conception.[7] Some methods alter the physical character of the act itself: the act is then deviant, i.e., intrinsically non-generative *qua* physical act. In other methods the physical nature of the act is not altered: but the act *qua* intentional act will be intrinsically non-generative. All forms of contraceptive sex are intrinsically non-generative *qua* intentional acts, and some are also intrinsically non-generative *qua* physical actions.

Safe period sex, deliberately chosen with the intention of avoiding conception, is not condemnable, provided, of course, there is good reason to avoid conception in the particular case. This is for two reasons: *qua* physical acts, safe period sex acts are normal, i.e., intrinsically generative even if not generative in fact; and *qua* intentional acts they are also intrinsically generative. For the intention to avoid conception is not embodied in the act, as it would be if the act itself had been interfered with. The intention to avoid conception is merely a *further intention.*

In HW Professor Anscombe writes:

> In considering an action, we need always to judge several things about ourselves. First: is the *sort* of act we contemplate doing something that's all right to do? Second: are our further or surrounding intentions all right? Third: is the spirit in which we do it all right? Contraceptive intercourse fails on the first count. . . . An act of intercourse at an infertile time, though, is a perfectly ordinary act of intercourse, and it will be bad, if it is bad, only on the second or third counts.

Before moving on to the answer given by Professor Anscombe to the second part of her central question I would like to mention some difficulties that arise in connection with the answer (as above) given to the first part of that question.

It is clearly possible to draw a distinction between an intention which goes to make up the essential nature of an act, and an intention which does not. Thus the intention to deprive someone of his goods is part of the essence of *fraud* but it is not part of the essence of *telling tall stories* although one may tell tall stories with the intention of depriving someone of his goods. It is part of the essence of *contraceptive sex* that there is an intention to avoid conception but it is not part of the essence of *safe period sex,* under *that* description, that there is an intention to avoid conception, although one may choose the safe period with the intention of avoiding conception. The distinction, then, is obvious enough: but it is not always obvious just how and when it applies. Furthermore it is not obvious what the distinction rests on.

In his reply to Professor Anscombe Peter Winch argued that acts of safe period sex if chosen because they are infertile will follow a pattern and involve a 'method' and for this reason are indistinguishable, *qua* intentional acts, from contraceptive sex acts. That is to say, he argued, in effect, that *safe period sex acts which are governed by a method and chosen because they are infertile*, do, under *that* description, embody an intention to avoid conception. Now it seems to me that there is a difference between using a pill and using a method. In the former case the intention to avoid conception is a cause or part-cause of the act's being infertile. In the latter case the intention to avoid conception is indeed present but not as cause or part-cause of the act's being infertile. For in this case the act is infertile already. Or to put the point in another way: in contraception the further circumstances which determine whether the act will be fertile or not include the intention as cause: in safe period sex those circumstances include a natural state of the body as cause; here the intention still appears, but not as cause.

If the occurrence of the intention (to avoid generation) as cause is sufficient reason to call an act *intrinsically non-generative qua intentional act*, then it is only consistent to conclude that the occurrence as cause of an intention to produce generation can make an act of deviant sex *intrinsically generative qua intentional act*. Thus in artificial insemination the act of deviant sex will be properly described as *intrinsically generative qua intentional act* and *intrinsically non-generative qua physical act*. This conclusion seems a little odd to me but perhaps it is all right. The fact is that nobody knows whether the occurrence of the intention as cause is in every case necessary, or sufficient, or both, or neither, to determine the intrinsic character of an intentional act. To settle this question would require a whole new theory of intention of a kind which we do not at present have.

Be that as it may, Professor Winch's point suggests another similar point. Even if it is wrong to say that individual acts of safe period sex embody the intention to avoid conception might one not correctly describe the whole programme of choosing only safe period acts as embodying that intention?

The intention here occurs, not indeed as a cause of the infertility of this or that act, but rather as the cause of the infertility of a particular couple, of a particular couple's sexual life. That a programme can be said to embody an intention even if the acts which go to make it up do not seems clear from the example of legislation. A government might for instance pass a number of laws relating to marriage, taxation, and the duties of Customs Officers, not one of which could be said to embody the intention of preventing immigration—nevertheless in some cases the whole programme might be correctly described as embodying that very intention. It might be *of the essence* of the programme taken as a whole that it is intended to prevent immigration.

It seems then that we are left with the following question: if an *action* is condemnable because it embodies a certain intention, will a *programme of behaviour* which embodies that same intention also be condemnable, and if not why not? At present I do not see any way of finding an answer to this question. Let us suppose, however, that this difficulty has been resolved, either by deciding that safe period sex does not constitute a programme embodying an intention to avoid generation, or by deciding that the embodiment of an intention in a programme does not necessarily mean that the programme is bad even if a single act which embodied that intention would necessarily be bad for that reason: suppose for the sake of argument that the difficulty has been resolved along some such lines. Is there not a further difficulty, as follows?

(1) *If* you have shown that the differentiation of acts into different types requires some reference to intentions; and
(2) *If* you have shown that there is a general distinction between intentions which are as it were part of the intrinsic character of an act and intentions which though present are not part of the intrinsic character of an act; and
(3) *If* you have shown (what is perhaps obvious anyway) that the intrinsic character of an act is one of the things which determines whether the act is condemnable or not; and
(4) *If* you have shown that contraceptive acts are intrinsically different *qua* intentional acts from acts of safe period sex;

you still have not shown that contraception is condemnable.

Firstly because two kinds of act can be as different as you like without it necessarily following that either is condemnable—it is the *kind* of difference that makes a difference. And secondly because the kind of difference which Professor Anscombe has located does not seem to justify the conclusion that contraception is condemnable. For the located difference is this. In safe period sex a non-condemnable intention (to avoid generation) is a further intention, and in contraception a non-condemnable intention (to avoid generation) is embodied in the act (that is, makes it what it is). But how can an act which is *otherwise all right* be condemned on account of its embodying, as part of its intrinsic character, an intention which is *also all right*?

A possible solution can be found in Professor Anscombe's answer to the second part of her central question. The second part of her central question was: 'Why is the characteristic which makes contraception condemnable a condemnable characteristic?' And the answer given is: 'A sexual act which is *qua* intentional act, intrinsically non-generative is a condemnable act because it is an act which illicitly interferes with the course of nature'. Now some people may object to this on the ground that (according to them) there can be no valid distinction made between licit and illicit interferences with the course of nature. There is not space in this paper to offer a general justification of the distinction and unfortunately Professor Anscombe says little on the subject herself. But perhaps the distinction can be seen to have application if we look at a few possible examples. I myself believe that shaving is a licit interference with nature, and that many forms of factory-farming are illicit interferences with the course of nature. To totally remove the smallpox virus from the face of the earth strikes me as a perfectly licit interference with the course of nature; to totally remove the whale or the dolphin or the horse from the face of the earth would strike me as a completely illicit interference with the course of nature. But what makes the difference? In my view many things make the difference. Consequences and motives are among the things that make the difference but they are not the only things that make the difference.

Let us assume then that a distinction *can* be drawn between licit and illicit interferences with the course of nature, and return to Professor Anscombe's claim that contraception is a case of illicit interference.

In CC Professor Anscombe writes: 'Disturbance of the order of nature may or may not be licit'. Thus eating (the action which is intrinsically nutritive) differs from sex (intrinsically generative) in that interfering with it, say for medical reasons, is not illicit. The Church condemns artificial insemination, but allows intravenous feeding; it forbids contraception, but allows dieters to swallow non-nutritive food-substitutes; and no doubt it would allow Professor Anscombe's drastic imagined case: '. . . to install a substitute for lung-breathing by some reversible operation (with a view to underwater exploration, say) . . . .'

So why is interference with one process (generation) forbidden, while interference with others (eating, breathing) is allowed? We might suppose that sex is more important than the other two functions because sex is what *transmits life*. But after all eating and breathing *preserve life*.

In HW Professor Anscombe says:

> There is no such thing as a casual, non-significant sex act. . . . Contrast sex with eating—you're strolling along a lane, you see a mushroom . . . you know about mushrooms, you pick it and you eat it quite casually—sex is never like that.

And in CC: 'Casual eating is harmless—you see a mushroom in a meadow as you walk by, and you pick it and eat it without shame or shamelessness'. She goes on:

> There is a deep association between sex and shame. No one will deny this: some may think it culture-bound, and that we should try to get away from it. But it is bound to too many cultures for that to be credible.

She argues that this shame is not a mark of disgracefulness—'it occurs as it were senselessly'. But 'the reasonless shame has to be respected'. A comparison is made between the reasonless shame associated with sex and the (reasonless?) respect that is felt towards a dead body. We know it would not be right to treat a dead human body casually, as something that could be stuffed into the garbage can for collection on Wednesday. Such knowledge, she says, is common to all humanity, and it is mystical. By 'mystical knowledge' she means partly a knowledge of what is right which has no utilitarian basis.

I do not know, of course, if shame about sex is found in all or most societies; but the proposition strikes me as very believable. And even if shame is not as common as we suppose, still it is surely the case that all human societies treat sex as something which is essentially significant and non-casual. For sex is everywhere controlled by laws and rules and complicated customs.

It is also the case that eating is sometimes non-casual, and that it is controlled by rules, the rules of etiquette and hospitality. Eating has connections with religion too—consider fasts and fasting, and the practice of saying grace before meat, and the food-sacrifices made to gods. One thing we can learn from the example of eating is that what can be casual needn't be casual always. And since sex *is* the activity which transmits life it certainly isn't going to turn into something that is *always* casual.

Rules about sex seem to have a more profound importance than the rules which surround eating and drinking. To call this 'mystical knowledge' may be right. But part of the explanation must surely be more mundane. The rules that govern sex link up—via birth, inheritance, and property—with the whole of the rest of the legal system. This must be so in any society which has property and which recognises kinship.

Must the 'reasonless shame' be respected? There is certainly no reason why it should not be respected, but perhaps we should bear in mind, as we concede our respect, that the reasonless shame has in fact produced unnecessary rules, unreasonable prohibitions, and a lot of lying. Unnecessary rules: for example, the rule that women must not wear men's clothing and vice versa. Unreasonable prohibitions: for example those which forbade women to study medicine (in case they saw naked bodies) or to use libraries (in case they disturbed the men). Lying: for example the lies once standardly told to children when replying to their questions about birth and procreation.

If we allow that we have a mystical knowledge that sex ought not to be casual, can we move from this knowledge, from this premise, to the conclusion that *a physically normal sex act which embodies a non-condemnable intention is itself condemnable on that account*? The step looks to me like a *non sequitur*; but I would not be so rash as to claim that Professor Anscombe will not on some later occasion fill in the gap and produce a watertight argument.

Notes

1 G. E. M. Anscombe, 'You can have sex without children: Christianity and the new offer', in *Renewal of Religious Structures: Proceedings of the Canadian Centenary Theological Congress* (Toronto, 1968).
2 *The Human World*, May 1972.
3 *The Human World*, November 1972.
4 G. E. M. Anscombe, *Contraception and Chastity* (Catholic Truth Society, London, 1977).
5 G. E. M. Anscombe, 'On *Humanae Vitae*'; paper presented to the First International Conference on the Regulation of Birth and the Ovulation Method of Natural Family Planning held at the University of Melbourne, 9–19 February 1978.
6 For instance, K. D. Whitehead, of Catholics United for the Faith Inc., New York, writes in *L'Osservatore Romano*: 'We have legalized abortion

today because we had social acceptance of contraception first', and speaks, in this connection, of 'the necessary link between contraception and abortion'.

7 The expression *embodies an intention* is mine, not Professor Anscombe's. But I believe it captures her thought.

III

SENSE AND NONSENSE

[11] A WITTGENSTEINIAN SEMANTICS FOR PROPOSITIONS

Bogusław Wolniewicz

Part I

More than once Professor Anscombe has expressed doubt concerning the semantic efficacy of the idea of an 'elementary proposition' as conceived in the *Tractatus*. Wittgenstein himself eventually discarded it, together with the whole philosophy of language of which it had been an essential part. None the less the idea is still with us, and it seems to cover theoretical potentialities yet to be explored. This paper is a tentative move in that direction.

According to Professor Anscombe,[1] Wittgenstein's 'elementary propositions' may be characterized by the following five theses:

(1) They are a class of mutually independent propositions.
(2) They are essentially positive.
(3) They are such that for each of them there are no two ways of being true or false, but only one.
(4) They are such that there is in them no distinction between an internal and an external negation.
(5) They are concatenations of names, which are absolutely simple signs.

We shall not investigate whether this is an adequate axiomatic for the notion under consideration. We suppose it is. In any case it is possible to modify it in one way or another, and for the resulting notion still to preserve a family resemblance with the original idea. One such modification is sketched out below.

Part II

Let us assume the reference of contingent propositions to be *possible situations*. This fundamental notion is really an offshoot of the correspondence theory of truth. For let α be any true proposition, and let the line R represent all *reality* in Wittgenstein's sense (i.e., the totality of facts) as shown in Fig. 11.1:

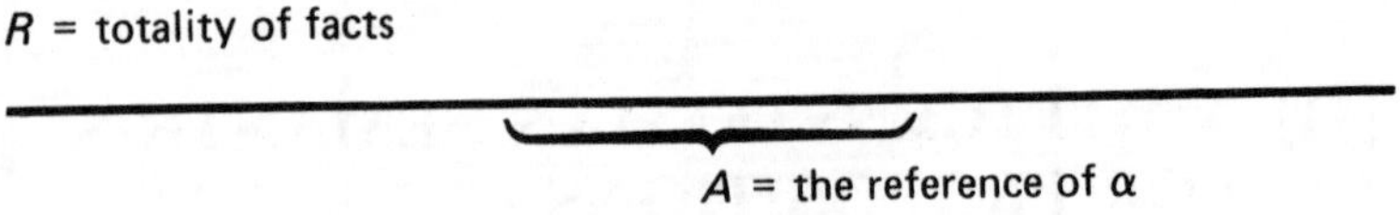

R = totality of facts

A = the reference of α

Fig. 11.1

Being true, α corresponds to reality, but not all reality is relevant to that. Consequently, R splits up into the segment A referred to by α, and into the vague remainder indifferent to it. Thus A represents here the smallest fragment of reality warranting the truth of α. This is the reference of α, but obviously its truth is warranted also by any fragment A' greater than A. In that case we shall say: α is *verified* by A'. And any fragment of reality fit to verify a proposition is to be called a *situation*.

This much is just common sense. The next step, however, is an extremely controversial one, for we expand now the notion of reference so as to cover false propositions as well. Since there are no *facts* (i.e., *real situations*), to correspond to them, we postulate to that purpose *imaginary* ones. Both are *possible*, and so the totality of facts is embedded in the totality of possibilities. This consists of all the situations which can be described in the language considered. In a Pickwickian sense we shall still say that a proposition α is verified by a possible situation A, but now that only means that if A were real, α *would* be true.

An imaginary situation is a non-being. Hence to admit them as the reference of false propositions is to infringe what Plato had called 'the ban of the great Parmenides': 'Keep your mind from this way of enquiry, for never will you show that non-being is'.[2] In this, however, we follow in the steps of the great Frege, whose minimal semantics for propositions still admits of two situations:[3] the one real (*das Wahre*), the other one imaginary (*das Falsche*). The former corresponds to 'the One' of Parmenides and to 'the totality of facts' of Wittgenstein; the latter obviously has no counterpart in Parmenides, and no clear-cut counterpart in Wittgenstein.

Part III

Two semantic relations between a proposition α and a situation A should be kept clearly apart:

A is presented by α

α is verified by A.

Generally these are incompatible, but in one extreme case they coincide: the greatest situation presented by α is identical with the smallest situation verifying α. Following Meinong we call this situation the *objective* of α, denoting it by '$S(\alpha)$'. Thus the objective of a proposition and its reference are two distinct concepts with the same extension: $A = S(\alpha)$. In view of this they may be used interchangeably.

The objectives of contingent propositions should satisfy at least these two conditions:

(C1) $S(\alpha) = S(\beta) \equiv \alpha \Leftrightarrow \beta$

(C2) $S(\alpha)$ is part of $S(\alpha \wedge \beta)$

with '$\Leftrightarrow$' denoting strict equivalence, and 'part' taken in its mereological sense. To comply with them we introduce the notion of an *elementary situation* (or 'E-situation' for short) defined for a certain kind of language. Such languages are to be called the *Wittgensteinian* ones.

A set of propositions L is a W-language if and only if there is a subset $E \subset L$ such that: (i) all the members of L are truth-functions of those in E; and (ii) the set E may be partitioned into a finite family of classes — the *logical dimensions* of L — of which the following holds: members of one dimension are mutually incompatible, and members of different dimensions are mutually independent — in fact, W-independent.[4] Moreover, in each dimension one E-proposition is true, and at least one is false. The members of E are the *elementary propositions* of L. A group of W-languages is composed of those satisfying the equality $E = E^+ \cup E^-$, where the members of E^+ are elementary in the more usual sense of being 'atomic', and E^- is equivalent to the set of their respective negations. Each dimension then consists of a pair of E-propositions, like 'The switch is on' and 'The switch is off'. (Observe that the set E^+

corresponds well to the specifications laid down in Professor Anscombe's five theses.)

An *E-conjunction* is the conjunction of an arbitrary number of *E*-propositions each taken from a different dimension. (In the limiting case it may be a single *E*-proposition.) The objective of an *E*-conjunction is an *E-situation*, and the totality of the *E*-situations—denoted by '*SE*'—is to be our universe of discourse. The objectives of single *E*-propositions are the smallest *E*-situations, and they correspond to Wittgenstein's *Sachverhalte*. The greatest *E*-situations are *possible worlds*, and the set of all possible worlds is the *logical space* of *L*.[5] We denote it by '*LS*', and obviously $LS \subset SE$.

The set *SE* is a partial algebra. The operation in *SE* is the *join* of *E*-situations, i.e., their mereological sum, to be written '$u = s; t$' for any $s, t, u \in SE$. Since the join is the counterpart of a conjunction of *E*-conjunctions, we assume it to be idempotent, and also—if feasible—commutative and associative. Thus the *E*-situations are partially ordered by the relation of mereological inclusion: $s \leq t \equiv \text{df } s; t = t$. The formula '$s \leq t$' will be read '*s* inheres in *t*'.

Part IV

The objectives of contingent propositions are sets of *E*-situations (i.e., proper subsets of *SE*). A proposition α may come true in many ways (see *Tractatus* 4.463 (a)), each such way being an *E*-situation and belonging to $S(\alpha)$. The objective $S(\alpha)$ is a unit set only if α is an *E*-conjunction: in particular, an *E*-proposition (see Anscombe's thesis (3)).

Not every proper subset of *SE*, however, is the objective of some proposition. In order to determine exactly the family of objectives, we introduce on the basis of *SE* four further notions, three by definition, one axiomatically. Thus for any sets $A, B \subset SE$ we define the *minimum* of an *SE*-set, the *product* of *SE*-sets, and a relation of *involvement* between them:

(D1) $s \in \text{Min}(A) \equiv s \in A \wedge \sim \bigvee_{t \in A} t < s$

(D2) $u \in A \cdot B \equiv \bigvee_{s \in A} \bigvee_{t \in B} u = s; t$

(D3) $A \text{ involves } B \equiv \bigwedge_{s \in A} \bigvee_{t \in B} t \leq s$

with '$t < s$' in (D1) to mean, as usual, '$t \leq s \wedge t \neq s$'. Note that for the *SE*-sets set-theoretical inclusion is a special case of involvement: $A \subset B \rightarrow A$ involves B, with '$\leq$' reduced to identity.

By a *verifier* of a proposition α we mean any *E*-situation s such that if s were real, it would make α true. The set of all the verifiers of α we denote by '$V(\alpha)$', and obviously $V(\alpha) \subset SE$ for any contingent α. We characterize this notion by the following nine axioms, with 'w_0' marking the *real world*, and '$\Rightarrow$' denoting strict implication:

$$\text{(A1)} \qquad \alpha \text{ is true} \equiv \bigvee_{s} (s \in V(\alpha) \wedge s \leq w_0)$$

$$\text{(A2)} \qquad \alpha \Rightarrow \beta \equiv V(\alpha) \subset V(\beta)$$

$$\text{(A3)} \qquad s \in V(\alpha) \rightarrow (s \leq t \rightarrow t \in V(\alpha))$$

$$\text{(A4)} \qquad V(\alpha \wedge \beta) = V(\alpha) \cap V(\beta)$$

$$\text{(A5)} \qquad V(\alpha \vee \beta) = V(\alpha) \cup V(\beta)$$

$$\text{(A6)} \qquad s \in V(\sim \alpha) \rightarrow \sim (s \in V(\alpha))$$

$$\text{(A7)} \qquad \bigwedge_{s \in V(\alpha)} \bigvee_{s' \leq s} \rightarrow s' \in \mathrm{Min}\,(V(\alpha))$$

$$\text{(A8)} \qquad u \in \mathrm{Min}\,(V(\alpha \wedge \beta)) \rightarrow \bigvee_{s \in \mathrm{Min}(V(\alpha))} \bigvee_{t \in \mathrm{Min}(V(\beta))} u = s; t;$$

$$\text{(A9)} \qquad \sim (s \in V(\alpha)) \rightarrow \bigvee_{t} (s \leq t \wedge t \in V(\sim \alpha)).$$

Of these, (A1) and (A2) are clearly just special ways of reading the usual definitions of truth and of entailment respectively.

Now we are finally in position to define accurately—for a *W*-language!—the notion of being the objective of a proposition. The objective of α is simply the set of all its minimal verifiers:

$$\text{(D4)} \qquad S(\alpha) = \mathrm{Min}\,(V(\alpha)).$$

We shall see, however, that the family of minimal *SE*-sets—i.e., such that $A = \mathrm{Min}\,(A)$—is still too wide to coincide with that of objectives.

Using (D4) one easily proves the theorem:

$$\text{(T1)} \qquad \alpha \Rightarrow \beta \equiv S(\alpha) \text{ involves } S(\beta).$$

Since for *SE*-sets the relation of involvement is reflexive and transitive, and since for minimal *SE*-sets it is also antisymmetric, we get as a corollary of (T1) the theorem:

$$\text{(T2)}\quad \alpha \Leftrightarrow \beta \equiv S(\alpha) = S(\beta).$$

Thus condition (C1) has been satisfied.

If 'is part of' is allowed to mean 'is involved in', then (C2) has been satisfied too. And if $\alpha \wedge \beta$ is an *E*-conjunction, this is actually the most natural reading of that phrase. For then both $S(\alpha \wedge \beta)$ and $S(\alpha)$ are unit sets, and thus for some $s, u \in SE$ we have $S(\alpha) = \{s\}, S(\alpha \wedge \beta) = \{u\}$. Obviously

$$\{s\} \text{ is involved in } \{u\} \equiv s \leq u,$$

and it may be doubted if one could satisfy (C2) any better.

Part V

The objectives of truth-functions have to be functions of the objectives of their arguments. Indeed, the objective of a *conjunction* is the minimum of the product of the objectives of its conjuncts:

$$\text{(T3)}\quad S(\alpha \wedge \beta) = \text{Min}\,(S(\alpha) \cdot S(\beta)).$$

We omit here all the proofs—they may be found in a recent article by the author[6]—but just observe that this holds: $V(\alpha) \cdot V(\beta) = V(\alpha) \cap V(\beta)$.

The objective of a *disjunction* is the minimum of the union of the objectives of its disjuncts:

$$\text{(T4)}\quad S(\alpha \vee \beta) = \text{Min}\,(S(\alpha) \cup S(\beta)).$$

Moreover, for any minimal *SE*-sets, the operations indicated have all the formal properties necessary: both are idempotent, commutative, and associative, and they are both distributive with respect to each other.

The operation corresponding to disjunction may be determined even more closely. In fact we have:

$$\text{(T5)}\quad S(\alpha \vee \beta) = [(S(\alpha) \cup S(\beta)) - S(\alpha \wedge \beta) \cup [S(\alpha) \cap S(\beta)].$$

If the disjuncts are incompatible, the last terms in (T5) disappear, and for that special case we get simply: $S(\alpha \vee \beta) = S(\alpha) \cup S(\beta)$. To see, however, the necessity of those terms in the general case, consider the following example. Let s_1, s_2, t_1,

t_2, u belong to the three different dimensions D_s, D_t, D_u respectively, and let us have $S(\alpha) = \{s_1, s_2, u\}$, and $S(\beta) = \{t_1, s_2; t_2, u\}$. Then by (T4) one should have: $S(\alpha \vee \beta) = \{s_1, s_2, t_1, u\}$. And indeed one has: $S(\alpha \wedge \beta) = \{s_1; t_1, s_2; t_1, s_2; t_2, u\}$, and $S(\alpha) \cap S(\beta) = \{u\}$.

These relationships are readily seen in Fig. 11.2

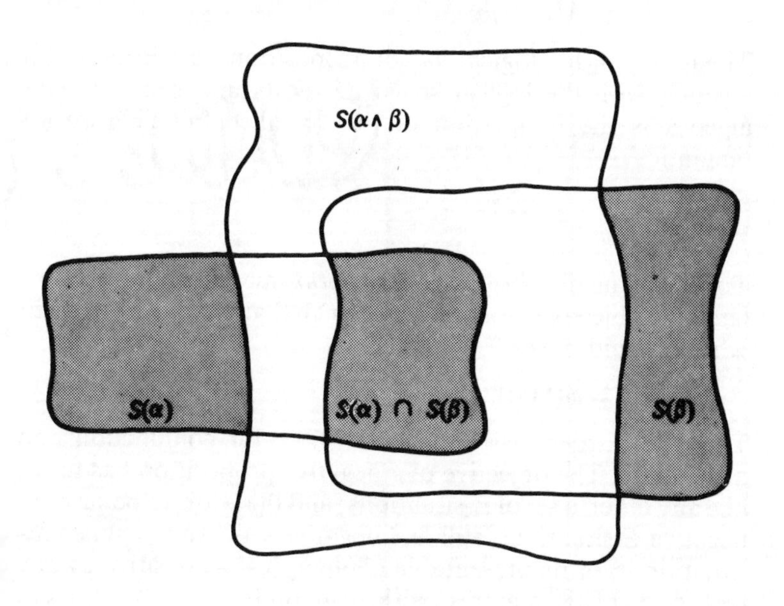

Fig 11.2

The shaded area corresponds to $S(\alpha \vee \beta)$; of course, $S(\alpha \wedge \beta)$ is *not* included in it.

Part VI

Before discussing the difficulties connected with the objective of negations, let us in passing consider another family of *SE*-sets. Having defined the *maximum* of an *SE*-set in analogy to (D1), we add an axiom analogous to (A7). Then we define the set $M(\alpha)$ of all the maximal verifers of the proposition α:

(D5) $M(\alpha) = \text{Max}\,(V(\alpha))$.

$M(\alpha)$ is a set of possible worlds, and a subset of LS, for we have: $LS = \text{Max}(SE)$. Thus $M(\alpha)$ is the *logical locus* of α, its '*logischer Ort*'.[7] Now stipulating $M'(\alpha) = LS - V(\alpha)$ we can prove the following theorems:

$$\text{(T6)} \quad \begin{cases} M(\alpha \wedge \beta) = M(\alpha) \cap M(\beta) \\ M(\alpha \vee \beta) = M(\alpha) \cup M(\beta) \\ M(\sim \alpha) = M'(\alpha). \end{cases}$$

The family of the logical *loci* of propositions is a Boolean set algebra, with the logical space LS as its unit element. This algebra is the foundation of the so-called 'possible-worlds semantics'.

Part VII

To determine the objective of a *negation* means to find—on the family of objectives—a function φ such that, for any contingent α, one should have:

$$S(\sim \alpha) = \varphi(S(\alpha)).$$

This is a harder task than were those for conjunction and disjunction. The objective of a negative proposition has to be, like any other, a set of E-situations; and there are, to be sure, no negative E-situations, still less negative sets. An imaginary E-situation is, in its presentable content, just as positive as any real one. (This squares with Anscombe's thesis (2).) On the other hand, two E-situations s and t may be incompatible, i.e., such that, for any $u \in SE$, we have $\sim (s \leq u \wedge t \leq u)$. Thus, in a way, we know quite well what $S(\sim \alpha)$ should be. For let us call a *falsifier* of α any E-situation verifying $\sim \alpha$. We denote the set of all falsifiers of α by '$F(\alpha)$', which is just short for '$V(\sim \alpha)$'. Hence, by (D4), the objective of a negation is the set of all the minimal falsifiers of the proposition negated: $S(\sim \alpha) = \text{Min}(V(\sim \alpha)) = \text{Min}(F(\alpha))$. This, however, does not give us yet any clue how to construct $S(\sim \alpha)$ out of $S(\alpha)$.

Every falsifier of α is incompatible with any of its verifiers: $V(\alpha) \cdot V(\sim \alpha) = \varnothing$. (By (A6), substituting $\sim \alpha/\beta$ in the theorem $V(\alpha) \cdot V(\beta) = V(\alpha) \cap V(\beta)$ mentioned before.) Consequently,

$$\text{(T7)} \quad S(\alpha) \cdot S(\sim \alpha) = \varnothing.$$

It follows immediately that $S(\alpha) \cap S(\sim\alpha) = \varnothing$. Note, however, that if α is *independent* of β (see *Tractatus* 5.152(a)), then the intersection of their objectives is empty too. (Though, of course, not that of their V-sets.)

Moreover, in a sense yet to be defined, $S(\alpha)$ and $S(\sim\alpha)$ have to *complement* each other. (Observe that $V(\alpha) \cup V(\sim\alpha) \neq SE$.) With that in view, we now introduce some further notions based on the universe SE.

Part VIII

Suppose the W-language L under consideration to have n logical dimensions, and let $D_1, D_2, \ldots, D_n$ be the sets of *Sachverhalte* corresponding to them. These are the dimensions of the logical space of L: $LS = D_1 \cdot D_2 \cdot \ldots \cdot D_n$. (This is not the Cartesian product!) Now we call the product Q_i of any k dimensions of LS, $1 \leq k \leq n$,

$$Q_i = D_{i_1} \cdot \mathrm{D}_{i_2} \cdot \ldots \cdot D_{ik}$$

a *quasi-subspace* of LS, or 'Q-space' for short. Evidently, there are $2^n - 1$ such Q-spaces, and they are all *disjoint*: if $i \neq j$, then $Q_i \cap Q_j = \varnothing$. The universe SE is the union of all Q-spaces:

$$\text{(T8)} \quad SE = \bigcup_i Q_i.$$

Each Q-space contains exactly one real E-situation, viz. the projection of w_0 upon Q_i. (An E-situation is real iff all its components are real, otherwise imaginary.)

We shall say, for any $A, B \subset SE$, that B is an *idle expansion* of A iff, for some Q_i, one has: $B = A \cdot Q_i$. This relation is transitive: if $B = A \cdot Q_i$ and $C = B \cdot Q_j$, then—by putting $Q_k = Q_i \cdot Q_j$, and since the product is associative—we also have $C = A \cdot Q_k$. Moreover, if $A \subset Q_i$, then evidently $A \;\; Q_i = A$. The idleness consists in this: if B is an idle expansion of A, then for any α, $A \subset V(\alpha) \equiv B \subset V(\alpha)$. Or to put it differently: if B is an idle expansion of A, then A contains some real E-situation iff B contains one. Now by a *quasi-complement* of an SE-set A we mean any set A' such that some idle expansion of their union is a Q-space; i.e., a set A' such that $(A \cup A') \cdot Q_i = Q_j$, for some Q_i, Q_j. If now the set A happens to be the objective of α, and

$A \cdot A' = \varnothing$, then the set Min (A') seems a plausible choice for the objective of the negation of α.

The quasi-complement, however, is not unique. For suppose LS has just two binary dimensions $D_s = \{s_1, s_2\}$, $D_t = \{t_1, t_2\}$; like {*switch on*, *switch off*}, {*door open*, *door shut*}. Hence $LS = D_s \cdot D_t = \{s_1; t_1,\ s_1; t_2,\ s_2; t_1,\ s_2; t_2\}$. Let α, β be the E^+-propositions presenting s_1, t_1 respectively, and let $\gamma = \alpha \wedge \beta$. Then $S(\alpha) = \{s_1\}$, $S(\beta) = \{t_1\}$, $S(\gamma) = C = \{s_1; t_1\}$; but what about $S({\sim}\gamma)$? It should be the set of all the minimal falsifiers of γ, i.e., $S({\sim}\gamma) = C'_1 = \{s_2, t_2\}$. And indeed, putting $Q_i = Q_j = LS$ we have $(C \cup C_1) \cdot Q_i = Q_j$. But the same goes for the sets $C'_2 = \{s_1; t_2,\ s_2\}$, $C'_3 = \{s_2; t_1,\ t_2\}$, and $C'_4 = \{s_1; t_2,\ s_2; t_1,\ s_2; t_2\}$, which are the *minima* of the idle expansions of C'_1.

Thus far we have got a function of $S(\alpha)$ determining only a whole family of SE-sets, of which the set $S({\sim}\alpha)$ is just one member. All members of that family, however, are equivalent in the following clear-cut sense: for any two of them, there is an SE-set which is the idle expansion of both; i.e.,

$$\text{(D6)} \quad |A| = |B| \equiv \bigvee_i \bigvee_j A \cdot Q_i = B \cdot Q_j.$$

Evidently, the relation defined in (D6) is reflexive and symmetric. It is also transitive, for if $A \cdot Q_i = B \cdot Q_j$, then again putting $Q_k = Q_i \cdot Q_j$ we have $A \cdot Q_k = B \cdot Q_k$. And if $B \cdot Q_1 = C \cdot Q_1$, then by further putting $Q_m = Q_k \cdot Q_1$ we get $A \cdot Q_m = C \cdot Q_m$. Hence the family of all the SE-sets—and thereby the family of all the *minima* of those sets too—is partitioned into the equivalence classes of that relation. (In our example we have, in particular, $|C'_1| = \ldots = |C'_4|$.)

If we restrict both the domain and the range of the relation of idle expansion to the family of the *minima* of SE-sets, it remains transitive, in view of the easily proved equality: Min $(A \cdot \text{Min}\,(B))$ = Min $(A \cdot B)$. But now it is also antisymmetric, i.e., if B = Min $(A \cdot Q_i)$ and A = Min $(B \cdot Q_j)$, then $A = B$. Anticipating here another formal move, which incidentally secures the reflexivity of idle expansion (see Part IX), we may say that the SE-sets in each of the equivalence classes defined are partially ordered by that relation. Now it is a fair conjecture that with respect to that ordering there is in each such class a unique *minimum*; i.e., an SE-set such that all the other members of that class are—in the sense of (D1)—the *minima* of its idle expansions. To distinguish this new *minimum* from the other one, let us denote it by '$\text{MIN}\,|X|$', unless the conjecture fails. Thus $A = \text{MIN}\,|X|$ iff $B \in |X| \equiv B = \text{Min}\,(A \cdot Q_i)$ for some Q_i. And if $A = \text{MIN}\,|A|$, we shall say that the *SE*-set A is dimensionally *irreducible.*

This much granted, the function φ is given by the following prescription for producing $S(\sim\alpha)$ out of $S(\alpha)$. Take any Q_i sufficiently high to make the set $A = S(\alpha) \cdot Q_i$ dimensionally homogeneous, i.e., such that $A \subset Q_j$, for some Q_j. Then take the minimum of the equivalence class of the difference $Q_j - A$. This will be $S(\sim\alpha)$. Or to put it in a formula:

$$S(\sim\alpha) = \text{MIN}\,|Q_j - S(\alpha) \cdot Q_i|,$$

where Q_i, Q_j have to be such that $S(\alpha) \cdot Q_i \subset Q_j$.

Part IX

Only contingent propositions have been considered so far, i.e., such that the set of their verifiers and the set of their falsifiers were both proper subsets of *SE*. Facing the question of the objective of a tautology, let us go back to the previous example. There we had $S(\alpha) = \{s_1\}$. Putting $Q_i = Q_j = D_s$, we have $S(\sim\alpha) = \text{MIN}\,|D_s - \{s_1\} \cdot D_s| = \{s_2\}$. On the other hand—by (T5), and $\alpha, \sim\alpha$ being incompatible—we have $S(\alpha \vee \sim\alpha) = S(\alpha) \cup S(\sim\alpha)$, and hence $S(\alpha \vee \sim\alpha) = \{s_1, s_2\} = D_s$. By the same token, however, we have also $S(\beta \vee \sim\beta) = D_t$, and so $S(\alpha \vee \sim\alpha) \neq S(\beta \vee \sim\beta)$, which violates our initial condition (C1).

This difficulty could be met in more than one way. We might for example make the condition (C1) more severe by interpreting '$\alpha \Leftrightarrow \beta$' not as the strict equivalence of C. I. Lewis, but as the 'strong equivalence' of A. A. Zinoviev, which obtains only if the sets of *E*-propositions occurring in α and β coincide. We prefer, however, to approach it in what seems to be more the spirit of the *Tractatus*.

We go now beyond the universe *SE* by adding to it the *empty E-situation*, denoted by '0'. In the wider algebra SE', $SE' = SE \cup \{0\}$, that situation is the neutral element characterized by the axiom: $s; 0 = s$, for any $s \in SE'$. It follows that $0 \leq s$, for any $s \in SE'$. Inhering in every *E*-situation, 0 inheres also in any real one. However, the components of a real *E*-situation are all real themselves, and so we have to assume that the empty *E*-situation is a real one too.

Evidently, in the family of *SE'*-sets the unit set $\{0\}$ is the neutral element with respect to the product: $A \cdot \{0\} = A$, for any $A \subset SE'$. Denoting it by 'Q_0', we regard it as another *Q*-space; the formula (T8) is extended appropriately. Geometrically *SE'* might be represented by a quarter of the plane together with the bounding half-lines. These, without their common end-point, represent the two proper *Q*-spaces Q_1 and Q_2. The topological interior of the quarter is the logical space $LS = Q_1 \quad Q_2$, with the point w_0 somewhere inside. The end-point—or rather the unit set comprising it—then represents Q_0.

In view of (D6) all the *Q*-spaces belong to the same equivalence class:

(T9) $|Q_i| = |Q_j|$, for any Q_i, Q_j.

Moreover, every *Q*-space is an idle expansion of Q_0, for taking $Q_i = Q_j$ we have $Q_j = Q_0 \cdot Q_i$ for any Q_j. Thus Q_0 is the smallest element in the class $|Q_j|$, i.e., $Q_0 = \text{MIN}\,|Q_j|$. And so we have:

$$\text{MIN}\,|S(\alpha \vee \sim \alpha)| = \text{MIN}\,|S(\beta \vee \sim \beta)| = \{0\}.$$

This is the objective of tautology: the 'void centre', *substanzloser Mittelpunkt* (see *Tractatus* 5.143(c)) of all the (objectives of) propositions. Thus if τ is a tautology, then $S(\tau) = \{0\}$. Hence $0 \in V(\tau)$, and so by (A3) we have $s \in V(\tau)$ for any s, i.e., $V(\tau) = SE'$. Consequently, $M(\tau) = \text{Max}\,(V(\tau)) = \text{Max}\,(SE') = LS$. Thus the *locus* of tautology is the whole logical space; which is just as it should be.

All this looks promising, but there are impediments ahead. One is the question of what to do about the objective of contradictions. Another is the fact that in the extended universe *SE'* the axiom (A5) is no longer valid: $V(\alpha) \cup V(\beta) \subset V(\alpha \vee \beta)$, but not conversely. There are other ones too. For any one of them there is some remedy, but perhaps there is no remedy for all.

Part X

Up to now no mention was made of *objects*, nor of how they

relate to the *situations*. This, of course, is a notoriously difficult question, and we have only a faint suggestion how it might be tackled. In the first place, we do not believe that it is profitable to start directly from individual objects, trying somehow to match them with single *E*-propositions. On the contrary, our point of departure would be the concept of the *substance of the world* (see *Tractatus* 2.021), i.e., of the totality of objects.

The world-substance *SW* is correlated with the logical space *LS* in that it is common to all possible worlds, i.e., to all the *E*-situations belonging to *LS*. Now according to the principle of Logical Atomism (see *Tractatus* 1.2–1.21) *LS* decomposes into a hierarchy of *Q*-spaces, many of them independent, though all variously involved in each other. Thus it seems reasonable to assume that there is a parallel decomposition of *SW*, and that for every *Q*-space Q_i there is a *sub-substance* SQ_i associated with it, the correlation $Q_i \leftrightarrow SQ_i$ being one–one. Therefore the question to raise here is how the totality of objects is distributed over the *Q*-spaces. To this question we find a hint of an answer in Frege.

Starting with *E*-situations, one arrives at objects by splitting the situations up in a Fregean way into function and argument. The argument is the object. For a given *E*-situation each such split brings forth other objects embedded in it.[8] There are as many objects embedded in an *E*-situation as there are splits of it. Thus with any *E*-situation *s* we may correlate the set $O(s)$ of all its objects. Given two different *E*-situations, it may then turn out that the same set of objects is correlated with them. Now it is a fair guess that all the *E*-situations correlated with the same set of objects $O(x)$ constitute a *Q*-space, $O(x)$ being its sub-substance. Hence any two co-dimensional *E*-situations would be two different configurations of the same objects.

Notes

1 G. E. M. Anscombe, *An Introduction to Wittgenstein's Tractatus*, (Hutchinson, London, 1959), chapter entitled 'Elementary propositions'.
2 Plato, *The Sophist*, 237A and 258D; Jowett's translation.
3 See R. Suszko, 'Ontology in the *Tractatus* of L. Wittgenstein', *Notre Dame Journal of Formal Logic*, 9(1968), 7–33.

4 See B. Wolniewicz, 'Four notions of independence', *Theoria*, 36 (1970), 161–164.
5 See E. Stenius, *Wittgenstein's 'Tractatus'*, (Basil Blackwell, Oxford 1960), chapter entitled 'Logical space'.
6 B. Wolniewicz, 'Situations as the reference of propositions', *Dialectics and Humanism*, 5(1978), 171–182.
7 See Stenius, *op. cit.*, *Tractatus* 3.4–3.42 and B. Wolniewicz, 'Zur Semantik des Satzkalküls: Frege und Wittgenstein'; in: *Der Mensch—Subjekt und Objekt. Festschrift für Adam Schaff* (Wien, 1973), pp. 411–419.
8 G. E. M. Anscombe, *op. cit.*, pp. 101–102.

[12] INTENSIONAL ISOMORPHISM AND NATURAL LANGUAGE SENTENCES[1]

Deirdre Wilson

It is well known that the linguistic form of a sentence may affect its semantic or pragmatic interpretation in extremely subtle and complex ways. Sentences such as (1) and (2), although logically equivalent, are generally felt to be semantically non-equivalent, and their semantic differences are generally attributed to their differences in linguistic form:

(1) War is war.

(2) A brother is a brother.

Certainly, utterances of (1) and (2) are not normally pragmatically equivalent: (1), but not (2), for example, might be considered a relevant answer to (3), while (2), but not (1), might be considered a relevant answer to (4):

(3) Why did the British bomb Dresden?

(4) Why did John Kennedy appoint Robert Kennedy Attorney General?

There are various ways of describing these facts: one would be to say that semantic and pragmatic interpretation consist not merely in establishing the entailments of sentences or utterances, but are also dependent on certain aspects of their linguistic form, so that (1) and (2), although logically equivalent, are none the less not semantically or pragmatically so. In that case, the task for semantic and pragmatic theory would be to say just how differences in linguistic form affect semantic and pragmatic interpretation.

From (1) and (2) it can be shown that any account of synonymy which appeals to mere logical equivalence is much too gross: it would predict as synonymous many sentences or utterances which are clearly not. However, once synonymy is allowed to become dependent on linguistic form, the danger is that the resulting accounts will be much too fine: they will fail to predict as synonymous many sentences or utterances which clearly are. For example, (5) and (6) differ in their syntactic structure, as do (7) and (8):

(5) He looked up the number.
(6) He looked the number up.
(7) Mary's friend is a bachelor.
(8) Mary's friend is an unmarried man.

Suppose one attempted to define synonymy as follows: two sentences (utterances) are synonymous if and only if they are logically equivalent and identical in syntactic and lexical structure. Then, on the face of it, (5) and (6) cannot be synonymous, nor can (7) and (8); in fact any two sentences which differ, however slightly, in syntactic form, will be automatically predicted as semantically non-equivalent.

In setting up his definition of intensional isomorphism, which is just such a syntax-dependent semantic concept, Carnap argues that certain syntactic differences between sentences are 'inessential', and semantically irrelevant, while others are essential and semantically relevant. Thus, for Carnap, (9) and (10) do not differ in essential syntactic ways, and are hence claimed to be intensionally isomorphic and synonymous:

(9) $2 + 5 > 3$.
(10) Greater [Sum (II, V), III].

On the other hand (11) and (12) differ in essential syntactic ways, and are therefore claimed to be neither intensionally isomorphic nor synonymous:

(11) $7 > 3$.
(12) Greater [Sum (II,V), III].

The crucial difference between (11) and (12) is that (11) contains the syntactically simple expression '7' where (12) contains the complex expression 'Sum (II,V)'. According to Carnap, this difference, unlike that between 'Sum (II,V)', and '2 + 5', has crucial effects on semantic interpretation.[2]

Two points about Carnap's account of intensional isomorphism are perhaps worth special emphasis. First, he is suggesting that linguistic form may affect semantic structure without affecting actual entailments expressed. One would like to know not just *that* linguistic form may have some effect on semantic structure, but also *what* effect it has. Second, he is providing for a distinction between essential and inessential differences in syntactic structure. We want to know not just that there is a distinction, but also how it is to be drawn. In particular, it seems that for Carnap, while (5) and (6) might be regarded as differing only inessentially, and hence still be analyzed as synonymous, (7) and (8) cannot be regarded as synonymous, because (7) contains the syntactically simple expression 'bachelor' while (8) contains the syntactically complex expression 'unmarried man'. Is there any genuine reason for drawing the distinction between essential and inessential syntactic differences along these lines? Are there, at least, pragmatic differences between (7) and (8) which might be traced back to this putative difference in their semantic structure?

When natural rather than artificial language sentences are taken into account, a third point also becomes relevant. Sentences which share their syntactic structure may none the less differ phonologically. (13) and (14) are examples, as are (15a–d):

(13) Your son has measles.
(14) Your son has rubeola.
(15a) BILL'S father writes books.
(15b) Bill's FATHER writes books.
(15c) Bill's father WRITES books.
(15d) Bill's father writes BOOKS.

From Carnap's account, it would seem that the phonological differences between (13) and (14) are inessential, and that this case merely reduces to that of (9) and (10). However, whether the same is true of (15a–d) is an open question. It is clear that (15a–d) are pragmatically appropriate to rather different contexts: for example (15b) might be an appropriate answer to (16a) or (16b), but not to (17a) or (17b):

(16a) Who writes books?
(16b) What relation of Bill's writes books?
(17a) What does Bill's father do?
(17b) What does Bill's father write?

(15d), on the other hand, might be an appropriate answer to (17a) or (17b), but not to (16a) or (16b). If these pragmatic differences are in turn attributable to differences in the semantic structures of (15a–d), then it seems that the phonological differences between these sentences, unlike those between (13) and (14), are essential. And in this case, in turn, one needs an account of essential versus inessential phonological differences, and of their effect on semantic structure.

Notice that because of the way the notion of intensional isomorphism was set up, it does not matter that (15a–d) are logical equivalents: they may still be semantically non-equivalent. In fact, I shall argue that every sentence, even with a given stress pattern and syntactic description, has a number of alternative, and non-equivalent, semantic descriptions, with associated differences in pragmatic interpretation. The resulting account of semantic structure—an account which involves a partial ordering of the entailments of a given sentence or utterance—seems to provide not only a characterization of intensional isomorphism for natural language sentences, but also a solution to a long-standing problem about the presuppositional interpretation of utterances.

1 Presuppositions and intensional isomorphism

In the last ten years or so, there has been a vast outpouring of linguistic literature attempting to account for the semantic or pragmatic interpretation of utterances in terms of a notion of presupposition. It has not been generally noticed that the presupposition problem and the problem of intensional isomorphism are quite intimately connected: both are concerned with the finer differences of interpretation induced by the linguistic form of sentences or utterances which are otherwise logically equivalent. For example, (18a) and (18b) are logically equivalent, but differ in their syntactic structure and pragmatic interpretation:

(18a) God exists and reigns in heaven.
(18b) God reigns in heaven.

For presuppositionalists, these two sentences differ in that (18a), by virtue of its linguistic form, entails (19), whereas (18b), by virtue of *its* linguistic form, presupposes it:

(19) God exists.

Presuppositions are also generally held to be sensitive to phonological form: thus, for presuppositionalists, (15b) might be analyzed as presupposing (20a) or (20b), whereas (15d) might be analyzed as presupposing (21a) or (21b):

(20a) Someone writes books.
(20b) Some relation of Bill's writes books.
(21a) Bill's father does something.
(21b) Bill's father writes something.

This immediately raises the question of whether the presupposition problem and the problem of intensional isomorphism are not reducible to the same thing, and whether a solution to one might not automatically provide a solution to the other. The presupposition problem is essentially that of accounting for the fact that not all the propositions in the interpretation of an utterance are equally relevant, informative or salient: the problem would be solved if it could be shown how the linguistic form of an utterance imposes a structure on the propositions it expresses, and how this structure is pragmatically interpreted in terms of degrees of relevance. The intensional isomorphism problem is essentially that of showing how two logically equivalent sentences which differ in their linguistic form may also differ in their semantic structure: a solution to the presupposition problem, on the lines just mentioned, would seem to provide an automatic solution to the problem of intensional isomorphism too.

In the past few years, a certain orthodoxy has grown up around the treatment of presuppositions in linguistics. The generally accepted view is that presuppositions must be either semantic, in which case they are defined in logical terms as truth-or-falsity conditions rather than more standard truth-conditions; or pragmatic, in which case they are preconditions on the appropriateness or relevance of utterances. Moreover, a number of quite convincing arguments have been given against the semantic treatment.[3] For example, while (22) and (23) have preferred interpretations on which (24) is also true, it is hard to see how (24) could be a logical presupposition of (22) and (23):

(22) It is possible that Bill's brother is a cheat.
(23) John says that Bill's brother is a cheat.
(24) Bill has a brother.

It is clearly possible for both (22) and (23) to be true even if (24) is false: a possibility which is ruled out if (24) is a logical presupposition of (22) and (23). For this reason, and others, the idea that semantic presuppositions have any central role to play in linguistic theory has been increasingly abandoned by presuppositionalists.

The only alternative has seemed to be to explain the preferred interpretations of (22) and (23), and other 'presupposition'-carrying utterances, in terms of a notion of pragmatic presupposition, defined as a precondition on the relevance or appropriateness of utterances. For a long time, one of the main defects of such treatments has been the lack of any explicit definition of relevance or appropriateness. However, recent work by Sperber and Wilson suggests that relevance may be defined along the following lines: an utterance is relevant in a given context to the extent that it has pragmatic implications in that context; where a pragmatic implication is defined as a proposition implied neither by the context alone, nor by the utterance in isolation from the context, but by the utterance and context combined. Such a definition would, at least in principle, remedy one of the main defects of the pragmatic approach to presupposition.[4]

However, there is a much more serious defect in this approach, and one that has not often been remarked on. Wherever the semantics–pragmatics borderline is drawn, it seems indisputable that certain aspects of utterance-interpretation fall on the pragmatic rather than the semantic side, and that these aspects of interpretation are governed by considerations of relevance. For example, while it is the task of semantics to state the range of possible interpretations of an ambiguous sentence, it is pragmatic theory that must account for the actual disambiguation of utterances in context. Thus (25) is ambiguous, because *play* may mean either *play an instrument* or *play a game*:

(25) Most boys play badly.

If (25) is said in response to (26), the interpretation will generally be that most boys play games badly; and the reason for this is that only on this interpretation can (25) be seen as a relevant response to (26):

(26) Most girls play games well.

Similarly, while the grammar may place restrictions on the range of *possible* referents of 'the king' in (27), it is clearly the task of pragmatic theory to describe how the actual referent is chosen when (27) is uttered in context:

(27) The king is dead.

The ascription of reference, again, seems to be determined by considerations of optimal relevance: reference will be ascribed in such a way that the resulting statement is the optimally relevant one in that context. Considerations of relevance also seem to govern the cases in which an utterance is interpreted as saying more than what the sentence uttered literally means. Conversational implicatures are the obvious examples; but consider also the case where (29) is an answer to (28), and would thus be naturally construed as in (30):

(28) Does John play the violin well?

(29) He plays very badly.

(30) John plays the violin very badly.

(29) does not entail (30), but if it is to be a relevant answer to (28), it will have to be interpreted along the lines of (30). Thus, in at least three central aspects of pragmatic interpretation, a principle of optimal relevance plays a crucial role.

It would be tempting at this point to make the very strong claim that *only* a principle of relevance is necessary for purposes of pragmatic interpretation. Or, in other words, that pragmatic interpretation can be explained in terms of a general definition of relevance, together with a description of sentence meaning and a description of context and speaker-hearers' assumptions about the world. It seems to me that the only bar to drawing such a conclusion about pragmatic theory is the claim that there are pragmatic presuppositions which have no prior basis in semantic structure. The effects of linguistic form on presuppositional interpretation are largely arbitrary, in the sense that differences in the interpretation of utterances with the same syntactic structure but different stress patterns, or the same truth-conditions but different syntactic or phonological structures, could not be predicted on the sole basis of a general definition of relevance. It seems that, for English, there are simply arbitrary rules which link differences in stress patterns and syntactic structure to differences in the possibilities of interpretation for a given sentence or utterance. It is the very arbitrariness of these rules which indicates that they cannot be pragmatic. And in this case, it seems that the purely pragmatic approach to the problem of presuppositional interpretation—like the earlier semantic approach—must be rejected as inadequate.

These two approaches have always seemed to presuppositionalists to exhaust the possibilities of accounting for presuppositional interpretation. There is, however, an alternative approach, which, because it is semantic, avoids the defects of the purely pragmatic treatment, but also avoids the defects of the logical-presuppositional approach. Instead of using two different formal types of truth-condition—presuppositions and entailments—to account for differences in semantic structure, one might instead attempt to analyze such semantic differences as resulting from differences in the organization of truth-conditions of a *single* formal type. Such an approach—which will be pursued in the next section—would avoid the problems inherent in the semantic presuppositional treatment, and would also provide for a much more highly structured account of sentence-semantics than the simple, two-level treatment implicit in the semantic presuppositional approach.

2 Ordered entailments

The syntactic structure of a sentence may be represented in the form of a tree, the syntactic structure of (31) being roughly that in (32).

(31) The boy left the house

(32)

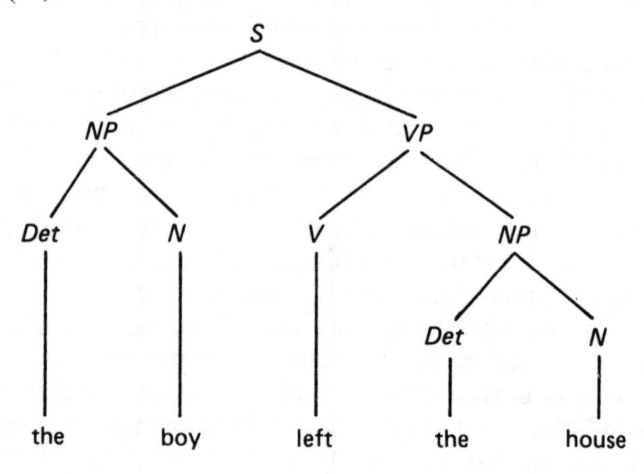

Fig. 12.1

(32) assigns each word in (31) to a syntactic category, and groups certain words into larger syntactic constituents. The following mechanical operation may be performed on (31): replace the first word by its category label as given in (32) and retain the resulting string; restore the first word and replace the second by its category label; restore the second word and replace the third, and so on until the end of the sentence. The resulting strings are given in (33a–e):

(33a) *Det* boy left the house.
(33b) The *N* left the house.
(33c) The boy *V* the house.
(33d) The boy left *Det* house.
(33e) The boy left the *N*.

The same operation may be performed on each of the larger syntactic constituents in (31): the resulting strings are given in (33f–i):

(33f) *NP* left the house.
(33g) The boy left *NP*.
(33h) The boy *VP*.
(33i) *S*.

The strings in (33) count as entailments of (31) if they are interpreted along the following lines. (33f), for example, is read as in (34):

(34) There is at least one noun-phrase of English which may be substituted for the *NP* label in (33f), the result being a true sentence.

On this interpretation, (31), with the syntactic description in (32), entails (33f); and by similar arguments, it entails all of (33a–i). Call (33a–i) the *grammatically specified entailments* of (31).

These entailments have an internal logical structure: some of them entail, or are entailed by, other members of the set. For example, (35) consists of a strictly ordered subset of (33a–i): a subset in which each member entails its successor and is entailed by its predecessor:

(35a) The boy left the house.
(35b) The boy left the *N*.
(35c) The boy left *NP*.
(35d) The boy *VP*.
(35e) *S*.

It is also a *maximal* strictly ordered subset, in the sense that no other member of (33) could be added to it, and the result still be strictly ordered. There are other such subsets of (33) such as (36a–d), for example:

(36a) The boy left the house.
(36b) The *N* left the house.
(36c) *NP* left the house.
(36d) *S*.

Call each such subset a possible *focal scale* of (31).

Within this framework, it is natural to see the stress-pattern of a sentence as determining the choice of an actual focal scale from the set of available possibilities, with different stress patterns determining different choices. If (31) is stressed as in (37), for example, it seems that the prescribed focal scale is that in (36):

(37) The BOY left the house.

If it is stressed as in (38), (35) is the prescribed focal scale:

(38) The boy left the HOUSE.

There is a straightforward method for determining the prescribed focal scale for a sentence with a given stress pattern: it is the scale containing the grammatically specified entailment in which the most heavily stressed word is replaced by its syntactic category-label: in which 'BOY' is replaced by *N* in (37), and 'HOUSE' by *N* in (38).

The stress-pattern and syntactic structure of a sentence do not, however, determine a unique semantic or pragmatic interpretation. It seems that the notion of a *focus* is necessary for this. The focus of a sentence is normally defined as one of the syntactic constituents which contain the most heavily stressed item, and in general there is more than one such constituent. Thus for (37) the focus might be the *N* 'boy', the *NP* 'the boy', or the sentence as a whole. For (38) it might be the *N* 'house', the *NP* 'the house', the *VP* 'left the house', or the sentence as a whole. Given the focus of the sentence, however, its semantic structure may be uniquely determined in the following way. Replace the focus by its syntactic category label; the result will be a grammatically specified entailment on the prescribed focal scale. Thus, suppose the *NP* 'the boy' is the focus of (37), its focal scale will be (36), and the result of substituting '*NP*' for 'the boy' yields the grammatically specified entailment (36c). Call this the *first background entailment* of (37) with focus 'the boy'; call the entailments above it on the focal scale [(36a) and (36b)] the *foreground entailments*, and those below it (36d) the *background entailments*. Then the choice of a focus for a given sentence divides its focal scale into two subsets, consisting of its foreground and background entailments, and should determine a unique pragmatic interpretation of the sentence in the following way.

The 'presuppositions' of a sentence with a given focus are its background entailments, including the first background; moreover, any grammatically unspecified entailment entailed by the first background may also exhibit presuppositional behaviour. Thus, consider (38) with first background (39):

(38) The boy left the HOUSE.

(39) The boy left the *N*.

(39) entails (40), which is standardly treated as a presupposition of (38):

(40) There was a boy.

(39) also entails (41), which is also presupposed by (38) on this interpretation:

(41) The boy did something.

In general, such entailments, including the background entailments, are pragmatically interpreted as preconditions on the relevance of an utterance of (38). Denying them amounts to denying the relevance of (38); questioning them amounts to questioning the relevance of (38), and so on. In this way the idea of a pragmatic presupposition as a necessary condition on relevance may be reconstructed—on the basis of a prior description of semantic structure which has no resort to any formal distinction between entailments and logical presuppositions.

By contrast, the foreground entailments of an utterance are pragmatically interpreted as determining the point of the utterance; to deny them is to deny the point of the utterance, rather than its relevance, and to question them is to question its point, rather than its relevance. Moreover, any grammatically unspecified entailment which entails the first background entailment may also form part of the point of the utterance. Consider (42), which is one such entailment of (38):

(42) The boy left the building.

(42) might well form part of the point of (38) on a particular occasion of utterance, and to deny it would be to deny that the utterance was true, but not that it was relevant.

Within this framework, a wide range of facts about the pragmatic interpretation of utterances may be explained on the basis of two principles: one arbitrary and semantic, and one general and pragmatic. The arbitrary principle states the effect of linguistic form on semantic structure, and is thus correctly located within semantics; the general principle states that degrees of semantic ordering are pragmatically interpreted in terms of degrees of relevance, and is thus correctly located within pragmatics. In this way, the defects of both logical and pragmatic presuppositional approaches are avoided, and some content has been given to the claim that two logically equivalent sentences which differ in their linguistic form may differ in both their semantic and their pragmatic interpretations.[5]

3 Intensional isomorphism and ordered entailments

Returning to the problem of intensional isomorphism within the framework just outlined, it is now possible to make a number of specific claims about the effect of linguistic form on semantic interpretation. First, synonymy for utterances may be defined as follows:

(43) Two utterances are synonymous if and only if they are logically equivalent and have the same focal scale, focus and background.

The definition in (43) makes the interesting prediction that only certain aspects of linguistic form are relevant to semantic interpretation. It predicts, for example, that two logically equivalent utterances will always be semantic (and hence pragmatic) equivalents *unless* their focuses differ in linguistic form. In other words, linguistic form is irrelevant to semantic interpretation except where it affects the choice or form of the focus. The resulting definition of synonymy is not so gross that it allows no aspect of linguistic form to affect semantic interpretation; but it is also not so fine that it requires every aspect of linguistic form to have some effect on semantic interpretation.

Returning to some earlier examples, consider (7) and (8) (repeated here for convenience):

(7) Mary's friend is a bachelor.

(8) Mary's friend is an unmarried man.

Suppose that the focus for both (7) and (8) is the *NP* 'Mary's friend', so that their backgrounds are (44) and (45) respectively:

(44) *NP* is a bachelor.

(45) *NP* is an unmarried man.

(44) and (45) are not, of course, sentences of English: for present purposes they may be regarded as expressing identical or equivalent propositions. In that case, (7) and (8) have the same focal scale, focus and background and are thus, by the definition in (43), synonymous. On the other hand, suppose that the focus of (7) is the *N* 'bachelor' and that of (8) is the *NP* 'Mary's friend'. The notion of focus is crucially dependent on linguistic form: two constituents which differ in their linguistic form to the extent of containing different words could never count as identical focuses. Hence, on this interpretation (7) and (8) do not have the same focus, and would not be predicted as synonymous. The same applies if the focus of (7) is the *NP* 'a bachelor' and that of (8) is the *NP* 'an unmarried man': again, (7) and (8) do not have the same focus, and would not be predicted as synonymous. There seems to be a certain amount of pragmatic confirmation for these predictions: two utterances which differ in their focus will normally be appropriate to rather different contexts, or differ in their pragmatic implications in the same context.

A similar point may be made with respect to (13) and (14):

(13) Your son has measles.

(14) Your son has rubeola.

If the focus is 'your son', or 'has', in both (13) and (14), they will be predicted as synonymous. However, if it is 'measles' in one and 'rubeola' in the other, then they will differ in focus and be predicted as non-synonymous on this interpretation. For Carnap, (13) and (14) would have to be treated as synonymous, because they are intensionally isomorphic. This has always raised problems for his account, because (46) and (47) would therefore, for him, also have to be synonymous:

(46) Anyone who believes that your son has measles

believes that your son has measles.

(47) Anyone who believes that your son has measles believes that your son has rubeola.

Yet there do seem to be certain circumstances in which (46) could be true and (47) false. The treatment just proposed suggests that there are certain interpretations on which they would not be synonymous, as well as interpretations on which they would mean the same.

Conclusion

My main purpose in this paper has been to show that the problem of intensional isomorphism and the problem of accounting for the presuppositional interpretation of utterances may be approached in the same way. Both problems result from the fact that the linguistic form of two logically equivalent sentences may affect their semantic (and pragmatic) interpretation. I have argued that an adequate account of intensional isomorphism for natural language sentences can provide an automatic solution to the presupposition problem; and conversely, that an adequate solution to the presupposition problem requires some reconstruction of a notion of intensional isomorphism; and I have given a brief sketch of one approach to handling both problems. My conception of intensional isomorphism is not, of course, a standard one. I hope I have not stretched it beyond all recognition.

Notes

1 The ideas put forward in this paper are based on work done jointly with Dan Sperber. I am grateful to him for permission to use them. I would also like to thank Ruth Kempson, Colin McGinn, Geoff Pullum and Neil Smith for comments on an earlier draft.

2 See Carnap *Meaning and Necessity* (University of Chicago Press 1956), pp. 56–7.

3 See for example, G. Gazdar *Formal Pragmatics for Natural Languages: Implicature, Presupposition and Logical Form* (unpublished Ph.D. thesis University of Reading 1976); Ruth Kempson *Presupposition and the Delimitation of Semantics* (Cambridge University Press 1975); Deirdre

Wilson *Presuppositions and Non-Truth-Conditional Semantics* (Academic Press 1975).

4 See Dan Sperber and Deirdre Wilson *Semantics, Pragmatics and Rhetoric* (forthcoming) for a more detailed account of relevance.

5 For further discussion of the views put forward in this section, see Sperber and Wilson (*op. cit.*) and Deirdre Wilson and Dan Sperber "Ordered entailments: An alternative to presuppositional theories", in C.-K. Oh and D. Dineen, eds. *Syntax and Semantics 11: Presuppositions* (Academic Press 1979).

[13] FREGE AND NONSENSE
Cora Diamond

O dieses ist das Tier, das es nicht gibt. (Rilke)

Frege says that it is impossible to assert of an object the sort of thing that can be asserted of a concept.[1] If, for example we were to take a sentence asserting of a concept that there is something falling under it (say, 'There is a King of England'), we may attempt to construct a corresponding sentence about an object; the result will be a sentence of sorts ('There is Queen Victoria'), but what corresponds to it will not be an illogical thought, but no thought. It will not say of the object named that it has something that as a matter of logic only concepts can have.—I shall raise some questions about Frege's view, with the aim of making clear what it involves and how it is related to the abandoning of a psychologistic account of meaning. The starting point for me was Professor Anscombe's illuminating discussion of Wittgenstein's later view of nonsense.[2] I wanted to see how far that view could be found in the *Tractatus*, and that led to the question whether its roots were not to be found in Frege.

Let us ask first whether it is possible to describe 'There is Queen Victoria' as a putting of an expression for an object—a proper name—where one for a concept should go. That is, is it possible to identify an expression as a proper name when it occurs in the wrong place, or what we want to call that? Frege thinks a proper name cannot be used predicatively; our question is whether, in a place where only a term used predicatively would make sense, we can identify a proper name at all. Take 'Chairman Mao is rare', an example of Michael Dummett's. Explaining Frege's views, he says that this is senseless because 'rare', although it looks like an ordinary adjective expressing a first-level concept, really expresses a second-level one—and so the idea is that we get a meaningless sentence when we put a proper name where the argument term should go. But suppose it were said that we do not always get nonsense that way: 'gold' Frege takes to be the proper name of an element,[3] and 'Gold is rare' is not senseless.—Well, whatever its role may be elsewhere, *there* it is not a proper name but means what not everything that glitters is. Frege himself points out that the same word can be used in some contexts as a proper name, and in others as a concept word; he says this in discussing an example somewhat like the present one.[4] But now, I can certainly say 'There isn't much Churchill in our present lot of politicians', and someone might equally use 'Chairman Mao' to mean a certain sort of political intelligence, which might indeed be rare. So how is it possible to tell whether, in 'Chairman Mao is rare', we have the proper name or a concept word? One indication Frege gives is this: if the expression in question occurs with an indefinite article or a numeral, or in the plural without an article, it is in that context an expression for a concept.[5]

But here we can see two problems. First, the shift of a word from proper name to concept word is exemplified for Frege by the shift from 'Vienna' the name of the city to 'Vienna' as used in 'Trieste is no Vienna'. Here, Frege says, it is a concept word, like 'metropolis'—but more specifically it is a word for a countable kind of thing. Shifts like that from 'gold' the proper name of an element to 'gold' used predicatively are not mentioned by Frege but are—it would seem—possible too. The occurrence of what is normally a proper name in a context in which an expression for a *countable* kind of thing would make sense (where this is shown by an indefinite article, the plural or a numeral) is taken by Frege—or so it seems—as a sufficient condition for treating the term as in that context a concept word and the occurrence as a predicative one; we must ask why the occurrence of what is normally a proper name in a context where a word for a *stuff* concept would make sense is *not* taken by Frege to be a sufficient condition for regarding the term as in such contexts a concept word. However, it now seems there may be a problem whether we can describe our original case as one in which the proper name 'Queen Victoria' has been put in the place where a concept word belongs. There will not be any examples of putting a proper name where a concept word belongs if the fact that a concept word belongs in a place is a sufficient condition for treating *whatever* is put there as in that context a concept word.

Let us look at Frege's own examples. He treats 'There is

Julius Caesar' as nonsense, but 'There is only one Vienna' not as nonsense but as a shift in the use of the word 'Vienna'.[6] What makes this possible is not that there is an established use of 'Vienna' as a concept word but that there is an established possibility *in the language* of using what are normally proper names as concept words. (If that were not so, the use of the indefinite article, the plural or a numeral could not be taken as Frege does: for example, 'As soon as a word is used with the indefinite article or in the plural without an article, it is a concept word.')[7] But now, what if we do not know what it is for something to be a Vienna? If this has not been settled, would the sentence still be describable as one in which 'Vienna' has the role of a concept word? If that were Frege's view, the question would be why 'There is Julius Caesar' should not be treated analogously. It seems it could be. For 'There is gold' ('Courage is rare') makes sense: why should we not say that 'There is Julius Caesar' ('Chairman Mao is rare') contains a concept expression whose reference has not been made clear? The first problem then with Frege's suggestions about how to tell when what is normally a proper name is being used as a concept word is that he does not treat in the same way all cases which could be regarded as involving a shift in the logical category of the term in question, and does not make clear the basis for this difference in treatment. (I do not mean that there is no basis. But I shall return to this question later.) The second problem arises immediately we take him to be allowing that 'There is only one Vienna' contains a concept word whose reference may be unclear. For if all cases apparently similar were treated in the way Frege treats that sentence, there would be no such thing as putting an expression into a place where an expression of a different logical category was required. There would instead, in all cases in which it was clear what logical category was required, be expressions whose logical category was clear from the context but whose reference might or might not be fully determined, expressions which in other contexts had a different categorial role and a fully determined reference. We could not then identify 'a proper name in the place where a concept word belongs' as a *proper name*; to speak of such an expression as a proper name at all would only be to refer to its role elsewhere, or to the role it was intended to serve. On this view, there would not merely be, as Frege clearly believed, no illogical thoughts (no such combinations of senses), but also no ill-formed sentences (no combinations of expressions violating categorial requirements), even in ordinary language—and this he clearly did not believe. In fact the possibility of ill-formed constructions in ordinary language is precisely one of the marks of distinction between it and Frege's symbolic language. And so it seems that if Frege takes a consistent view we have not yet found what it is.

The argument just given overlooks something. It *is* true that Frege believed that incorrectly formed names (including here sentences) were possible in ordinary language. But what is meant by 'incorrect formation' is less clear; the problem is that we may see in Frege a way of understanding 'incorrect formation' which is not there. Frege explains the correct formation of a name this way in the *Grundgesetze*, when he is stating the requirement for his symbolic language, that correctly formed names must always have a reference: 'A name is correctly formed if it consists only of signs introduced as primitive or by definition, and if these sign are used only as what they were introduced as being: thus proper names as proper names, names of first level functions of one argument as names of functions of this kind and so on . . . .'[8] It is clear from this passage that we can use such terms as 'proper name' to speak of the role an expression was introduced to serve, or to speak of its use in a particular context, the role it serves there. If the reference of a name in its intended use is fixed, but it is then used in some context in an entirely different way, it will be possible to form complex expressions containing it whose reference will not be determined, and it is this which Frege is concerned to avoid. We *can* speak of such uses of terms as 'violations of category requirements' but that expression is itself ambiguous. We may mean that an expression introduced as a term of one sort has been used with a different role, but a 'violation of category requirements' in this sense will not necessarily be nonsense: 'Trieste is no Vienna' is of this sort. To respect category requirements in this sense is simply to avoid cross-category equivocation. But we may also speak of a 'violation of category requirements' without suggesting that any term has been used in some other role than that for which it was introduced. That is, we may want to say that, *given* the role of the terms in some combination, the combination is nonsense—meaning that the trouble arises precisely *because* they have those roles: it is the fact that they have the roles, say, of a proper name and a second-level predicate, that prevents their being joined together to produce anything but nonsense. Here we have implicitly a sense of 'incorrect formation' different from that explained in the *Grundgesetze* passage, and a different notion of the violation of category requirements. The idea is intuitively of a 'clash' of the category of the terms combined, and it thus depends on the possibility of identifying the categorial role of a term outwith the context of legitimate combination. In particular, it assumes the possibility of identifying the categorial role of terms in a context in which (putting this in Fregean language) the reference of the parts does *not* determine the reference of the whole. In contrast, the *Grundgesetze* account does not assume that it is possible to identify the categorial role of a term, to say it has a sense of such-and-such a logical sort, in an incorrectly formed expression—except in the 'Trieste is no Vienna' sort of case, in which all that is necessary to assure a reference for the whole is that a reference for the parts be settled. The fact that Frege did believe that there could be ill-formed sentences in ordinary language does not then settle the question whether he thought there could in ordinary language be sentences which violated category requirements in the second sense (clashes of category as opposed to category-equivocations). The very questionable assumption on which that second sense depends was, I believe, rejected by Frege—and the rejection of it marks an important continuity in his philosophy of language, and between his and Wittgenstein's.

I shall discuss three points bearing on the question of Frege's view; two are considerations counting for and one apparently counting against what I take him to mean. First, that Frege does not think there can occur category violations in the second sense, even in ordinary language, fits well with the account he gives of the justification of a *Begriffschrift*, of its advantages over ordinary language. In *Über die wissenschaftliche Berechtigung einer Begriffschrift*, he puts great emphasis on the presence in ordinary language of equivocation—equivocation of a particularly dangerous sort if we are to think clearly: the use of the same word to symbolize both a concept and an object falling under it. A main advantage of a *Begriffschrift* is the rigorous avoidance of such equivocation. There is no mention of the presence in ordinary language of category error in the second sense, nor of its absence in a *Begriffschrift*.[9]

Secondly, and more important, the view I am ascribing to Frege is closely related to the *Grundlagen* point that we must ask for the meaning of a word only in the context of a sentence. The force of that principle is not clear, but it can be explained in terms of the notion of a logical part, a notion Frege contrasts with that of simplicity in *Grundgesetze* § 66.

> Any symbol or word can indeed be regarded as consisting of parts; but we do not deny its simplicity unless, given the general rules of grammar, or of the symbolism, the reference of the whole would follow from the reference of the parts, and these parts occur also in other combinations and are treated as independent signs with a reference of their own. In this sense, then, we may say: the word (symbol) that is defined must be simple.[10]

If we take a logical part of an expression to be one on whose reference the reference of the whole depends, in accordance with the general rules of the symbolism, we shall be making use of the notion of reference which Frege developed only after the *Grundlagen*. The notion of a logical part can also be explained in terms of the *Grundlagen* notion of content: a logical part of a sentence would be one on whose content the content of the whole depends in accordance with the general rules of the symbolism. 'Content' does not here correspond either to 'sense' or to 'reference'. When used of a concept word, relational term or proper name, it is close to the later notion of *reference*, but the content of a *whole* sentence, a 'judgeable content', corresponds neither to the truth-value of a sentence nor to its sense; it is much more like a Russellian proposition than is anything in Frege's later thought. Although this notion was not ultimately a useful one, it gives us as much as we need here: a reference-like notion, such that the whatever-it-is of the logical parts of a sentence determines the whatever-it-is of the whole.

Frege's original point, that we must ask for the meaning of a word only in the context of a sentence, reflects the idea that properly speaking it is only as a logical part of a sentence that the word has such-and-such a meaning; apart from such an occurrence we cannot even say that it is a concept word or a proper name. Thus take 'Vienna', which we may suppose to have been introduced as the proper name of a city. When it occurs in 'Trieste is no Vienna', the rules of English grammar make the content of the sentence depend on *what it is to be a Vienna*—and that is why we can say that it there appears as a concept word, and that it stands for whatever it is to be a Vienna. It may be totally undetermined what it is to be a Vienna, and in that case 'Vienna' would still be a logical part (because the rules of the language make the content of the sentence depend on whatever concept it has as content)—but it would then lack content, and so would the sentence. We may contrast the occurrence of 'Vienna' in 'Schlick founded the Vienna Circle'. Here it is not a logical part at all, and we cannot ask after *its* meaning in the strict sense.

On this view, to give the logical role of an expression in a sentence is to characterize the way the general rules of the symbolism determine the content of the whole sentence: if they do it via whatever item of such-and-such a sort (concept, say, or object or relation) is the content of *any* expression of some determinate pattern in that place, then the role of the expression is to stand for that sort of item. The expression occurs, for example, as a proper name if it occurs in a place where the content of the whole sentence depends on what object the expression there stands for. It cannot thus be identified as a proper name, or be said, strictly speaking, to have such a sense, if the content of the whole sentence does not depend on what object it stands for—in other words, if it occurs anywhere but in a suitable place. Nonsense of the category-clash type would then not be a possibility, even in ordinary language.

I have taken Frege's point to be that we must not ask, for example, of 'Vienna' what *its* meaning is, not taking it as a logical part of any sentence. He said that to ask for the meaning of a word in isolation led easily to a psychologistic account of meaning—and we can see one way it does so if we consider any word which can be used in more than one role; 'Vienna' will do.

We suppose it to occur apart from any context of which we can say something like this: the content of the whole depends on whatever such-and-such (concept, object or relation) 'Vienna' has as its content. And suppose we think of its occurrence as that of *a proper name*. That it *can* occur as a proper name does not account for its being one here in isolation; for it can also occur as a concept word. If we want to say it is one rather than the other, it is almost inevitable that we turn to the realm of the psychological for some connexion between *this* use of the word and the possibility it has of standing for a particular city in ordinary sentences. Perhaps it is a matter of my thinking of the city as I say the word. But it is not in virtue of any such psychological connexion that the word is a proper name in an ordinary sentence—and I may think of the city Vienna all I like when I say 'Trieste is no Vienna' and it will not on that account be a proper name there. We do not improve matters by saying I may describe 'Vienna' as referring to the city if I think of the class of sentences in which it does so. Thinking of a use does no more than thinking of a city to justify the ascription of a reference to a term in isolation. The same kind of appeal to the realm of the psychological is necessary if we think we can identify 'Vienna' as a proper name in a context of supposed 'category-clash'. *Ex hypothesi*, the content of the whole combination of words does not depend on the object 'Vienna' stands for. In virtue of what then is it supposed to be a proper name there?—What I am claiming is that exactly the reasons Frege had for saying that we should not ask for the meaning of a word in isolation are reasons for thinking the very idea of a category clash is confused.

These arguments depend on the way I have explained the *Grundlagen* point, and it may be said that Frege's insight should better have been put this way: that it is because an expression *can* occur as a logical part of sentences that we can unconfusedly ask for its meaning. That would allow for the significant occurrence, for example, of proper names on their own as greetings and so on—and would also allow the possibility of clashes of category. Simple versions of such a view may involve an implicit appeal to a psychological account of meaning in the way I have just sketched. However, I shall postpone discussion of the matter and instead ask what happens to Frege's insight if we treat sentences in the way he came to do, as merely a case of complete names.

One not unnatural move to try is to replace the *Grundlagen* principle with: Ask for the reference of a word only in the context of a completed function expression. Here then the logical parts of completed function expressions might have reference in the strict sense, and the categorial role of such a part would depend on the kind of rule-governed contribution made by its reference to the reference of the whole. What about completed function expressions themselves? Are there any conditions that would have to be met for us to ask for *their* reference? In fact, if we say that only logical parts of such expressions have reference, we shall already have included all complete names on their own. Any Fregean proper name, including ordinary proper names, would count as a case both of a completed function expression and of a logical part of a completed function expression: it can be regarded as a completion of an expression for the identity function by itself.[11]—The difference then between the earlier and the later views lies in part in where we can identify logical parts; they agree, though, in making the identification of the logical role of an expression depend upon the possibility of taking its occurrence as that of a logical part. And they agree in making an extremely close connexion between the possibility of unconfusedly ascribing a reference to an expression and its occurrence in a context with a certain sort of logical complexity—it is the characterization of this complexity which shifts.

I can now turn back to the question I raised earlier, about whether we should take Frege's insight to be that the essential thing for a term to have reference is the *possibility* of occurrence in a sentence as a logical part. The attractiveness of reformulating it in some such way is that (a) it allows for the ascription of reference to a proper name on its own, for example in greetings (as the original version of Frege's principle does not, but the revised version I just gave does), and (b) it makes clear the special significance of occurrence in *sentences* (as the original version does, but the revised version does not). But if it is stated as simply as I have done, it loses altogether a central feature of *both* versions, that it is only to an expression occurring in a context with a certain kind of logical complexity that reference can be ascribed. The use of an ordinary proper name in greeting someone, for example, could be taken to have the requisite complexity. *Who* it is one greets does not depend on the reference of the expression used in the greeting—when someone mistakenly uses my sister's name in greeting me, it is still I that am greeted. But what we might call the correctness or incorrectness of a greeting does depend on the reference of the expression used, and substituting terms with the *same* reference in the context of greeting someone or something will keep the 'correctness-value' of that greeting constant. In part, then, because it is a function of the reference of the expression used in greeting, it can be treated for certain purposes as the 'reference' of the greeting.

It is not my purpose to develop an account of the use of proper names on their own, but simply to suggest that accounts are possible which would not at the same time allow for 'category clash'. What leaves open the possibility of 'category clash' is taking as a sufficient condition for a word to have a reference that it *can* be used in a sentence as a logical part; but there are good reasons for wanting to avoid so weak a condition independently of the question about category clash. In any case, I am arguing now not that Frege was *right* in restricting reference to the logical parts of expressions with the kind of complexity exhibited by sentences (whatever exactly that kind of complexity is) and to what itself has such complexity, but that in so restricting reference, he was himself committed to a 'no category clash' view.

I have given two considerations in favour of thinking Frege did not allow the possibility of category clashes. But there is something apparently on the opposite side to which I shall now turn. For a long time, I thought that Frege's treatment of 'There is Julius Caesar' showed that he did not take the view I have been suggesting is his. For I thought the alternatives were only (a) that 'There is Julius Caesar' contains what is elsewhere a proper name, used as a concept word, and hence that it is no more senseless than 'Trieste is no Vienna', and (b) that it contains the proper name 'Julius Caesar' where a concept word ought to go, and is senseless for that reason. As Frege was clearly not taking the first alternative, he had to be read as taking the second or something like it, and hence as allowing the possibility of category-clashes. But that was wrong.

What Frege actually says about 'There is Julius Caesar' is simply that it is senseless. But what kind of senselessness is it? To answer that, I must explain why Frege takes such things as the presence of the indefinite article to indicate that a word is being used as a concept word. We may note that in the *Grundgesetze*, Frege says that we may form the name of a function (including here concepts and relations) by removing from a complex proper name (including here sentences) either a proper name forming a part of it (or coinciding with it) or a function name forming a part of it—in such a way that the argument place remains recognizable as capable of being filled by a name of the same sort as the one we removed.[12] The expression for a function is not recognizable as such without its argument place—and this is true even of ordinary language. Frege's *Grundgesetze* statement implies that we have not picked out a function expression unless we have associated with it a way of recognizing completions of that function expression as such. It is in part by means of the articles, numerals, and so on of ordinary language that we do so. Thus take:

(i) Anastasia lives in Charlottesville.
(ii) Another Anastasia lives in Charlottesville.
(iii) Another the King of France lives in Charlottesville.

Here (i) may be regarded as the completion of a first-level concept expression by the proper name of the last Tsar's daughter; (ii) cannot be taken to be the completion of *that* expression by 'Another Anastasia'. That is, the concept expression completed by 'Anastasia' in (i) can be recognized there because what precedes 'lives in Charlottesville' in the sentence is a noun in the singular on its own with *no* article, numeral, etc. The point is not that we can recognize the concept expression because there is a *proper name* before it. For any singular noun placed before 'lives in Charlottesville' on its own with no article etc. has one of the syntactical markers by which we tell that a term is being used as a proper name. I am not saying: 'any such noun in the argument place will be a proper name', because the argument place is not as it were a *place* at all. It is a place *for* a proper name or a bound individual variable, and if it has not got in it what has the syntactical marks of use-as-a-proper-name or use-as-a-bound-individual-variable, the 'place' simply is not there to be seen. That is, we cannot look to see what sort of expression is in the place where an expression for an argument ought to go: we can look at a combination of words to see whether it can be construed as such-and-such a concept expression *with* its argument place. If we have not got the syntactical markers of the argument place, we have not got the argument place *or* the concept expression.

On this account, (ii) cannot be taken to be the first-level concept expression '*x* lives in Charlottesville' completed by an argument expression, since a singular noun with 'Another' does not have the syntactical marks of use as a proper name. (It is not enough to point out that it *does* have the marks of use as a concept word; as we shall see, the syntactical markers of different sorts of terms in ordinary language may not be entirely distinct.) '*x* lives in Charlottesville' occurs in (ii) as part of the second-level concept expression 'Another (   ) lives in Charlottesville', which is completed by 'Anastasia', here a first-level concept word. (To be an Anastasia might be to be one who claims to be the Tsar's daughter Anastasia.) There is simply no such thing as putting the proper name 'Anastasia' which we see in (i) into the argument place of the second-level concept expression in (ii). I am not saying that any word or phrase that fills the argument place filled by 'Anastasia' in (i) would be used as an expression for a concept if we put it between 'Another' and 'lives in Charlottesville.' Frege's view is that 'the King of France' could fill the argument place filled by 'Anastasia' in (i), but it is not, in (iii), in the argument place of 'Another (   ) lives in Charlottesville'. In (iii), that is, we do not have an expression of the wrong sort in the gap filled by 'Anastasia' in (ii); the gap filled by 'Anastasia' in (ii) is simply not recognizable in (iii). The concept expression 'Another (   ) lives in Charlottesville' has, written into it, the means of recognition of the argument place—not through what kind of thing the expression there must stand for, but, with complete generality, what sort of expression it must look like. Whatever expression is there of that pattern, the reference of the whole will depend on the first-level concept it stands for—and that is what it is for it to be a term for such an item. On this view, then, (iii) does not contain any expression *in a gap in* another: it is simply a mess.

We can now note an ambiguity in the whole of the earlier part of this paper, which could not have been made clear earlier: I have spoken often of a 'place' where a term of a certain sort 'belongs', but what it is to 'belong in a place' is two distinct things. In (iii), we can turn the whole combination into sense by putting a singular noun where 'the King of France' is. In *a* sense, then, we can say that 'the King of France' does not belong there, and that it is its being there that makes nonsense of the whole. But it is not the case that 'the King of France' does not belong there on account of its logical category and the logical category of the rest. In (iii), what is left when 'the King of France' is removed has *no* logical category: it is not a concept expression of any sort and is not 'incomplete' in Frege's sense. In (ii), on the other hand, an expression for a first-level concept *belongs* in the argument place of 'Another ( ) lives in Charlottesville'. This is very different from saying that if such an expression is put there, the whole will make sense. If a singular noun is put there, or certain noun phrases, these expressions *will be* first-level concept expressions; that gives their use. If certain other noun phrases are put there, we shall not be able to identify the argument place and we shall not have put into a place requiring a first-level concept expression something that does not belong there. So in the sense in which we can say that a first-level concept expression belongs in a place, there is no method of identifying the place where it belongs, such that the place would *still* be identifiable if what were there could not itself be identified as a first-level concept term or appropriate variable.

In a *Begriffschrift* there will be completely unequivocal ways of making an argument place recognizable—but not so in ordinary language. That is, although it is possible (on the view I am ascribing to Frege) to make clear how the argument place of any concept or relational expression is indicated, even in ordinary language, there is no guarantee that argument places of fundamentally different kinds will always be marked in distinct ways. Frege himself points out, for example, that the (German) singular definite article does not always indicate that the noun which follows it stands for an object, and he suggests that what the plural definite article indicates is even less capable of simple formulation in a rule.[13]

Dummett has claimed that natural language constantly violates Frege's principle that a function expression cannot occur without its argument place or places—and he takes this to be consistent with Frege's principle.[14] I have been taking an opposite view: that the principle is not compatible with the identification of a function expression in natural language except *with* its argument place or places: that whatever the general form of the syntactical indicators of its argument places may be, that general form is part of the expression for the function. Natural language is *untidy* in its argument place indicators—hence the untidiness, and indeed incompleteness, of Frege's remarks about the definite and indefinite articles, numerals and plurals. A full account of the argument place indicators of, for example, the second-level concept expression we see in 'There is a book' would be quite long, and there was no reason for Frege to mention more than a few of its most conspicuous items.

How does this explain what Frege says about 'There is Julius Caesar'? Let me summarize my argument: Frege mistakenly thought that the only sort of argument place ever marked by the occurrence of a noun in the singular without an article or numeral or other explicit indicator is that of a first-level concept or relational expression. The second-level concept expression we can recognize in 'There is a horse' cannot therefore be recognized if we replace 'a horse' by a noun with what we might call the null indicator, like 'Julius Caesar'. It can also be argued that we cannot take the sentence as a first-level concept expression completed by 'Julius Caesar'. If it cannot be described as the completion of any expression by another—as a first-level concept term with a proper name, or as a second-level concept term with a first-level one, or in any other way—it is nothing but a word-hash, in no way logically different from a mere string of randomly chosen words. No logical role can be assigned to any of its parts, which are not logically parts.

What I shall argue more fully is the claim that Frege failed to recognize that the null indicator with a singular noun *may* mark a predicative use. In German as in English, a null indicator with a singular noun does normally show that the noun *as thus used* does not form a plural; it may, though, form a plural in other uses. Take 'brandy' as an example:

(i) Brandy has a higher alcoholic content than wine.
(ii) There is brandy but no soda.
(iii) That drink he finished in such a hurry was a brandy.
(iv) Cognac is a brandy and so is Armagnac.

'Brandy' as used here in (iii) and (iv) is predicative and forms a plural; as used in (i) it is (at least arguably) a proper name and does not form a plural. ('Brandies have a higher alcoholic content than wine' contains 'brandy' as used in (iv), not as used in (i).) As used in (ii), 'brandy' does not form a plural—and yet, I should say, it is as obviously a predicative use as 'There are horses' is of 'horses'. Why should there appear to be any problem about this? What exactly is the connexion between the capacity to form a plural and a predicative use—if, as I should say, that is at the root of Frege's blind spot about cases like (ii)? Frege does allow that not all concepts determine a principle for counting what falls under them, but when he discusses such cases he has in mind adjectives like 'red' and not stuff words.[15] What he says about 'red' is important: 'to a concept of this kind no finite number will belong.' His idea can be explained if we imagine constructing a list. If *A*, say, is some apple and is red, we may start our list with it, and if *B* is red and is not *A* (is a different apple or not an apple), we add it. (It might be the surface of *A*, or some particular patch on it.) I may go on with the list (making use each time of the criteria of identity associated with the names already on the list)—and suppose I am asked, when I have ten things on it, how many distinct things are red. I can say that *at least ten* are—but it is evident that a list so begun has no end. The significant thing for us is the contrast Frege wanted to make between on the one hand count nouns and adjectives, and on the other stuff words. I can have a horse, and *another* horse; something red (the apple), and something *else* which is red (its surface). It is just this feature which stuff words lack, on Frege's view. When I have told you that what is in *this* bottle is brandy, and what is in *that* one is brandy, I have given you the proper name of a substance, twice—*the* substance which is in both bottles. (Compare 'This plant is *Cassiope hypnoides*'—the 'is' is not the copula.) There is not an 'another' in the offing (except 'another bottle of brandy', but Frege would say that all that that gives us is a predicate of which the proper name 'brandy' is part); we have not got *two* things both of which are brandy, as we may have two apples or two things, however miscellaneous, both of which are red.— This account rests on Frege's 'Gold and gold and gold is never anything else but gold', a remark made in characterizing the use of a proper name as such.[16]

If I am correct that that was Frege's view of stuff words, what is wrong with it is that Frege has made too close a connexion between (a) the possibility of ascribing a predicate to one thing and then to *another* (where possibility means grammatical possibility, not necessarily freedom from contradiction) and (b) the possibility of forming a plural from the predicate, either directly ('apples') or, if the term is an adjective, through attachment to a plural noun. But when I say 'This is brandy and so is that', there *are* two quite different things that are said to be brandy, even though there are not two brandies nor two brandy things: the three ounces of liquid in here, and the twelve ounces of liquid over there—and the 'is' *is* the mere copula.[17]

The point, though, is not whether the view I have ascribed to Frege about stuff words is correct, but whether it is plausible to ascribe it to him. If it was his view, that would explain why he did not treat 'There is Julius Caesar' as he treated 'There is only one Vienna'. If singular-noun-with-null-indicator never indicates a predicative use, 'There is Julius Caesar' cannot be taken to contain the second-level concept expression we have in 'There is a horse' completed by 'Julius Caesar' used as a stuff word with no clear sense or reference. It would not contain the second-level concept expression *at all*—that is, the attempt to say of Julius Caesar what 'There is a horse' says of a concept does not *even* succeed in putting together expressions whose senses are unsuited to each other.

There are other passages in Frege—the last but one paragraph in 'On concept and object' is a good example—which may seem to count against the view I have ascribed to him, and which (with the 'There is Julius Caesar' passage) originally led me to think he *did* allow for the possibility of clash of categories. But rather than discuss in more detail how such passages should be taken, I shall consider two views closely related to the 'no-category-clash' view.

(i) There is *not*, in addition to the sort of nonsense we may produce by using a word or words with no determinate meaning, and the sort of nonsense we may produce by failing to adhere to the syntactical rules of ordinary language, another (and more philosophically interesting) sort of nonsense we may produce by putting together certain combinations of words violating no syntactical rule of ordinary language and using no word without a meaning. I shall call this putative sort of nonsense 'w-f', for 'well-formed'—well-formed, supposedly, in ordinary language, though ill-formed in some deeper way.

(ii) It is *not* a difference between ordinary language and a *Begriffschrift*, an adequate conceptual notation, that we cannot form in it sentences that correspond to the w-f nonsense of ordinary language.

Here (ii) is clearly a consequence of (i), and (i) is a consequence of the *Grundlagen* principle that a word has meaning only in a sentence. Thus, to take an example from a philosopher who makes use of a somewhat different system of logical categories from Frege's, there is Carnap's claim that 'Caesar is a prime number' is nonsense though syntactically well formed and without any meaningless words: he gives it as an example of what I am calling w-f nonsense.[18] Carnap believes '*x* is a prime number' is predicable only of numbers, which is to say in part that the truth or falsity of sentences resulting from completion of that expression by a name depends on what *number* the name stands for. But—applying a version of the Frege principle—if the truth or falsity of some sentence containing a name '*a*' depends on what such-and-such '*a*' is a name of, the role of '*a*' in *that* sentence is that of the name of a such-and-such. So in Carnap's sentence, 'Caesar' is the name of a number, and since it is not determined *what* number it is the name of, 'Caesar is a prime number' is not w-f nonsense after all, because it contains a number word with no determinate meaning.

I want to suggest that belief in w-f nonsense shows either failure to keep in mind possibilities of cross-category equivocation or an appeal to a psychologistic account of meaning. A clear case of the latter is Carnap's discussion of Heidegger on Nothing.[19] Carnap here too wants to persuade us that we have nonsensical sentences that break no rules of ordinary syntax and contain no meaningless words, and so he must dispose of the idea that Heidegger has, in saying all those weird and wonderful things about Nothing, simply departed from the ordinary meaning of the word. Now, prior to the sentences Carnap objects to, Heidegger has a sentence in which 'nothing' is used in the ordinary way. This shows, Carnap says, that in the passage as a whole, we should take the word 'nothing' to have the usual meaning of a logical particle serving for the formulation of negative existential statements. So when Heidegger says that 'the Nothing is prior to the Not' and what not, we are to take 'Nothing' as having the same meaning it has in 'There is nothing outside'. But what on earth *is* it for it to have the same meaning there? There is a gross misconception here of the role that can be given to a writer's intentions in settling what he means. One can no more look to a previous sentence to determine whether 'Nothing' is used as a logical particle in *this* one than one can look in one sentence to tell whether 'is' is the copula in another.

That belief in w-f nonsense may reflect failure to pay adequate attention to the possibilities of cross-category equivocation is illustrated by Dummett's discussion of 'Chairman Mao is rare', which he uses as an example of what I have called w-f nonsense.[20] The source of the difficulty here is not in any direct appeal to a psychologistic account of meaning. Rather, Dummett's claim is that 'rare' has the sense of a second-level predicate—that is a matter of the rules of the language. What then of its role in the sentence? Dummett takes the sense of a word in a sentence to be fixed by the general rules determining the sense of the word independently of any context, except in cases of ambiguity, where the rules of the language are not themselves sufficient to determine the sense the word has in particular sentences.[21] The kind of case he has in mind is that in which we have to guess from the context what the sense of a term is; but the sort of cross-category equivocation we have been concerned with is entirely different. In a case like 'Trieste is no Vienna' there is no question of *guessing* from the context or anything else what the role of 'Vienna' is. Further, we can recognize such cross-category equivocation even when the term in question has not antecedently been given two senses. That is, even if 'Vienna' has only been given the sense of a proper name, the last word in 'Trieste is no Vienna' is not a word whose sense is that of a proper name, occurring with the wrong sort of role or in the wrong sort of place. That word, there, is a concept word, and has, on our hypothesis, no specified sense as such. It has no logical relationship to the proper name 'Vienna'. The rule determining the sense of the proper name, the only sense determined for the word by the rules of the language, has no bearing on the sentence.

The case of Chairman Mao's rarity is similar. The *word* 'rare' has, in a sense, no sense. 'Rare' the second-level concept word has a sense, 'rare' the first-level concept word has several, and there is no such thing as combining the second-level concept word 'rare' with a term with the sense of a proper name. For the sake of simplicity let us forget, as Dummett does, that 'rare' has actually got several uses as a first-level concept word (including the 'O rare Ben Jonson' use), and let us ignore the 'Gold is rare' sort of case, too. Assuming, then, that 'rare' has no other use in the language than that of second-level concept word, we still cannot without further ado take the sense it has as such to have any logical connexion with the *word* 'rare' in some sentence—and in particular we cannot assume it to have any connexion with the word as it occurs in some nonsense sentence. In 'Chairman Mao is rare' we can in fact recognize the second-level concept expression 'Chairman Mao is ( )' completed by 'rare', which is there recognizable as a first-level concept word (it occurs in the argument place of a second-level concept expression with appropriate syntactical markers) as easily as the first-level concept word 'Shirley Temple' is recognizable in 'I'm going to have a blow-out on a Shirley Temple'.[22] We need not know what these first-level concept words mean in order to recognize them as such in these occurrences. If 'Chairman Mao is rare' is to be taken as w-f nonsense, we need to think that the sense both of 'rare' and of 'Chairman Mao', determined by the rules of the language (taking the one as second-level concept word and the other as proper name), have *some* connexion with that sentence. But the rules determining the sense of those two terms in that way cannot *both* be brought to bear on that sentence. And whether or not the word 'rare' has an established use as first-level concept word, the sentence can be taken as containing the word used in that way.

What then of the idea that an important part of the point of a good symbolic notation is that it excludes the w-f nonsense of ordinary language? Many philosophers have held such a view—and it has been ascribed to both Frege and Wittgenstein. But its ascription to them is more an indication of the hold the idea is capable of exerting on our minds than of any of their actual views. To make the issue here clear, consider the 'translation' into a symbolic notation of 'Trieste is no Vienna'. To be able to write this in a symbolic notation, the essential thing we need to know is that 'Vienna' is there a first-level concept word. The important difference between ordinary language and a *Begriffschrift* in the treatment of 'Trieste is no Vienna' comes out only when we put alongside the two versions, ordinary language and *Begriffschrift*, of 'Trieste is no Vienna' a sentence of ordinary language containing the proper name 'Vienna' used as such and a *Begriffschrift* version of *that* sentence. 'Trieste is no Vienna' in *Begriffschrift* differs from 'Trieste is no Vienna' in English in not containing anything that looks like the term that stands for the city in the other sentence. What is excluded from a *Begriffschrift* is only misleading appearances. It is not possible to take a sentence which is syntactically all right in English, and say that something corresponding to it will be excluded from a good notation.[23] Whatever it is, it will go over as easily as 'Trieste is no Vienna', but its *Begriffschrift* equivalent will lack any resemblance to the *Begriffschrift* equivalents of those sentences of ordinary language which had only superficial resemblances to the original sentence. The opposite view is a hangover of the kind of thinking most clearly exemplified by Carnap's idea that something logical particle-ish adheres to 'Nothing' in 'The Nothing is prior to the Not'. (Compare the idea in much *Tractatus* exegesis that something formal concept-ish adheres to 'object' in '*A* is an object'—which is then taken to be a sentence which could not be rewritten in a good symbolic notation.)

I want now to bring out more sharply the contrast between Frege's approach and the idea that the w-f nonsense of ordinary language is excluded in a *Begriffschrift*. The view I have ascribed to Frege implies that there are two ways in which an expression may be identified as a part of (or 'coinciding with') another:

(i) It is recognizable as in the argument place of some expression of level *n*, and is thus being used as one of level *n*-1.

(ii) It is an incomplete expression of level *n*, recognizable syntactically as completed by an expression or expressions used as expressions of level *n*-1.

(For the sake of simplicity I have omitted from both (i) and (ii) the case of unequal-levelled functions.)

The first method is the only one available for proper names: that is, there is no way of identifying the occurrence of a proper name in a sentence or in any other context except in the argument place of an incomplete expression of first level (including the identity function). This is a part of what Frege meant when he spoke of the decomposition of a sentence into a 'saturated' and 'unsaturated' part as a logically primitive phenomenon, which must simply be accepted and cannot be reduced to anything simpler.[24] Thus, although Frege himself tells us that we can form incomplete expressions by removing one expression from another, for example a proper name from a sentence, we get an incomplete expression as a result only if what we remove was *in an argument place*. We can, for example, distinguish the kind of pattern there is in 'Smith has Bright's disease', 'Smith has Parkinson's disease', 'Smith has Hansen's disease' from that in 'Smith has Bright's hat', 'Smith has Parkinson's hat', 'Smith has Hansen's hat' by recognizing that only in the latter set are 'Bright', 'Parkinson' and 'Hansen' in an argument place.

I have been arguing that it is not possible on Frege's view to identify the parts of a sentence or other complex expression independently of each other as expressions with certain logical powers. A complete knowledge of the sense or reference or both of all the expressions forming a sentence is not what enables us to recognize them in the context, since what has sense and reference is only expressions recognizable through function-argument decomposition as having a certain role in the context. But now, take the opposite view and ascribe it to Frege: We *can* identify the parts of a sentence independently of each other as expressions with certain logical powers, which may then be appropriately *or* inappropriately combined. A sentence that makes sense will result only if we combine parts whose logical powers fit each other, and any other combinations will yield nonsense. If we know the rules fixing the sense of an unambiguous term, we can recognize it in any sentence in which it occurs and grasp its contribution to it.—If we understand Frege that way, we shall find it impossible to take him literally when he says that an incomplete expression cannot occur without its argument places. This has to be interpreted to mean: *in an adequate notation*, an incomplete expression cannot occur without its argument places. Exactly the same happens with Wittgenstein. The idea that we can identify the role of an expression in a sentence via its role in the language, independently of the function-argument structure of the context in which it occurs, makes it impossible to believe he meant to be taken literally when he said that there is no need for a theory of types because 'what seem to be *different kinds of things* are symbolized by different kinds of symbols which *cannot* possibly be substituted in one another's places'.[25] These plain (I should have thought) words are taken to express the view that *in a correct notation*, different kinds of things are symbolized by different kinds of symbols which cannot possibly be substituted in one another's places. In the case of both Frege and Wittgenstein, statements about what is impossible in *any* language are taken to be about what is impossible in a special notation.

The discovery that, although we can put words together so that they make no sense, there is no such thing as putting together words with a certain role in the language, or with certain logical powers, so that on account of these roles or these powers, the whole is nonsense—this is surely one of the great things in Frege, and one of the most important things owed to him by Wittgenstein.[26]

Notes

1 G. Frege, 'On concept and object', in *Translations from the*

Philosophical Writings of Gottlob Frege, ed. P. T. Geach and M. Black (Basil Blackwell, Oxford, 1966), p. 50.

2 G. E. M. Anscombe, 'The reality of the past', in *Philosophical Analysis*, ed. Max Black (Cornell University Press, Ithaca, 1950), pp. 52–56.

3 G. Frege, *Foundations of Arithmetic*, trans J. L. Austin (Basil Blackwell, Oxford, 1974), pp. 49–50.

4 'On concept and object', p. 50.

5 *Foundations of Arithmetic*, p. 64, taken with 'On concept and object', p. 50.

6 'On concept and object', p. 50.

7 *Foundations of Arithmetic*, p. 64.

8 G. Frege, *Basic Laws of Arithmetic*, ed. M. Furth (California University Press, Berkeley and Los Angeles, 1967), § 28, p. 83.

9 G. Frege, 'On the scientific justification of a concept-script', *Mind*, 73 (1964), pp. 155–160.

10 *Translations*, p. 171.

11 *Basic Laws of Arithmetic*, §26, p. 81. See also P. T. Geach, 'Saying and showing in Frege and Wittgenstein', in *Essays on Wittgenstein in Honour of G. H. von Wright*, ed. J. Hintikka (North Holland, Amsterdam, 1977), pp. 59–60.

12 G. Frege, *ibid.*

13 'On concept and object', p. 45.

14 M. Dummett, *Frege, Philosophy of Language* (Duckworth, London, 1973), pp. 50–51, 178–179. Dummett's point is not easy to make out. He maintains in both passages that ordinary language does not adhere to the restriction that an incomplete expression can occur only with its argument place or places, but he also asserts that an expression can be identified as one with a certain sort of incompleteness only *with* its argument places. But whatever it is that is incomplete (for no word or words needs completion), *that* cannot appear without its argument place in ordinary language or anywhere else, on Frege's view. Dummett treats ordinary adjectives in attributive position as cases of expressions which are incomplete in virtue of their sense occurring without their argument places—which makes it seem as if the *word* 'blue' (say) has a sense of a certain sort which then requires it (and this requirement is then not adhered to in ordinary language) to come accompanied by an argument place of a certain sort. But what has such a sense cannot be identified with something that can occur without its argument place. The troubles here are those discussed by P. T. Geach in 'Names and identity', in *Mind and Language*, ed. S. Guttenplan (Clarendon Press, Oxford, 1975), pp. 147–50.

15 *Foundations of Arithmetic*, p. 66. See also P. T. Geach, *Reference and Generality* (Cornell University Press, Ithaca, 1968), pp. 38–39 and G. E. M. Anscombe and P. T. Geach, *Three Philosophers* (Basil Blackwell, Oxford, 1963), p. 86.

16 G. Frege, *ibid.* p. 50.

17 See P. T. Geach, *Reference and Generality*, pp. 39–40.

18 R. Carnap, 'The elimination of metaphysics through logical analysis of

language', in *Logical Positivism*, ed. A. J. Ayer (Allen & Unwin, London, 1959), pp. 67–68.

19 R. Carnap, *ibid.* pp. 69–71.

20 M. Dummett, *Frege, Philosophy of Language,* p. 51.

21 M. Dummett, *op. cit.* p. 268.

22 *The New Yorker*, 3 July 1978, p. 74. It is mostly ginger ale.

23 We should note that on Frege's view, a sentence like 'The moon is divisible by 2' fails to express a thought, not because it is illegitimately constructed but because no adequate definition has been given of '*x* is divisible by *y*'. A corresponding sentence is constructible in a good notation, and if the principles of definition are adhered to, it will express a thought. Again, we have in ordinary language, on Frege's view, sequences of grammatical sentences no one of which expresses a thought on its own. The whole sequence of such 'pseudo-sentences' does express a thought, and a single sentence of a *Begriffschrift* would correspond to it. See G. Frege, 'Über die Grundlagen der Geometrie', 1906, Part II, in *On the Foundations of Geometry and Formal Theories of Arithmetic*, ed. E-H. W. Kluge (Yale University Press, New Haven, 1971). pp. 69–103.

24 G. Frege, 'Über die Grundlagen der Geometrie', 1903, Part II, in Kluge, *op. cit.*, p. 33.

25 Letter to Russell, January 1913, *Notebooks 1914–1916* (Basil Blackwell, Oxford 1961), p. 121.

26 I am indebted to Peter Geach and Glen Kessler for helpful comments on a draft of this paper.

IV

TIME, TRUTH AND NECESSITY

[14] KINDS OF STATEMENT

Peter Geach

In its grammatical sense, as contrasted with 'question', 'command', etc., the term 'statement' is fairly unproblematic; it would be difficult to give a formal definition, and no doubt there are borderline cases, but well-educated people would find a very high measure of agreement about how the term applies to the sentences of a language. As with other grammatical terms, however, this measure of understanding does not involve an equal understanding of the logical role played by the linguistic entities to which the term is applied, and it is not surprising that we do not find one logical role played by all statements.

There is a prominent role played by very many grammatical statements: that of an asserted proposition. In this paper, as in much of my writing, I shall use 'proposition' in the sense, inherited from medieval logic, of a bit of language in a certain logically recognizable employment; long before Frege, and without commitment to his ontology of *Gedanken*, propositions were described as expressing complete thoughts, and accordingly being bearers of truth and falsehood. It is often said nowadays that truth-value appertains to a sentence only as 'used to make a statement', i.e. used assertorically; but this dictum is manifestly incorrect. Truth values belong, for example, to the two sides of a disjunction, and the disjunction's truth-value depends on those of its disjuncts, even though neither disjunct by itself is in this context being 'used to make a statement'. Etymologically 'proposition' suggests something propounded or put forward; to any such thing a truth-value already belongs, regardless of whether the proponent is or is not prepared to stick his neck out, to make a statement, when he propounds it.

An asserted proposition, then, may elsewhere occur unasserted without changing its sense or losing its truth-value. Recognizing repeated occurrences of the same proposition is not merely mechanical; the identity of a proposition is not the identity of a string of words. The proposition

'Socrates was bald' occurs over again in 'Socrates, who taught Plato, was bald', but does not occur in 'A philosopher whose teacher was Socrates was bald'. The like would hold for the corresponding Latin or Polish sentences; in Latin or Polish translation of the third English sentence just quoted, the proper name would be in the nominative case, but no proposition answering to 'Socrates was bald' would occur, although there would be the right words in the right order. Again, in the German sentences 'Die Erde ist rund' and 'Wenn die Erde rund ist, so ist die Erde rund', we have three occurrences of the same proposition, once asserted, twice unasserted, and the changes of word-order required by German idiom do not go against its identity.

I need not discuss here the many tricky questions and awkward cases that arise over recognizing a proposition's repeated occurrences in some vernacular. Even when we have not one and the same string of words, very often we can pick out one and the same proposition; in particular, we can often do this when the proposition occurs now asserted, now unasserted. A very frequent use of statements (I shall use this term henceforth in its grammatical sense, except where I explicitly indicate the contrary or am discussing authors who use 'statement' otherwise) is to propound propositions to whose assertion the speaker or writer is committing himself: but to these propositions it is not essential to occur in the guise of statements, for they are still recognizable in non-assertive occurrences.

This use of the statement form is so frequent and typical that I hope it will not be held tendentious to speak of *normal* statements. In recent philosophy a great deal of attention has been paid to statements that are, in this sense, not normal statements, or at least are alleged not to be. Writings on this matter are a favourite *genre* in the journals. The more radical theories of this *genre* will pick out some class of statements and argue that here no propositions are being asserted and no truth-value is assignable. A less radical type of theory holds that even between statements possessing truth-value there are radical differences; statements of different type are so to say logically insulated from each other; premises of type *A* never warrant conclusions of type *B*, nor premises of type *B* conclusions of type *A*. From this it is an easy step to the view that truth-values have to be ascribed in quite a different way when statements of different type are concerned.

Theories of this *genre* ought, I think, to be regarded with initial suspicion rather than initial favour. There certainly are non-normal statements; but many theories that make out some familiar class of statements to be not merely non-normal but also lacking in truth-value are very ill founded, and lose all plausibility when once we test them on examples outside the narrow range to which supporters of the theories confine themselves. Theories that divide true/false statements into logically insulated classes seem to me to be in far worse case: they are demonstrably false. There is, to my mind, no ground whatsoever for supposing that different kinds of statement have different kinds of truth; truth is just truth.

In what I have written thus far, and in the remainder of this article, I take for granted *the Frege point*, that a proposition can occur now asserted, now unasserted, without losing its identity or truth-value. I have said elsewhere, and see no reason to change my view, that only prejudice or confusion prevents people from seeing the Frege point, or allows them to maintain theories inconsistent with its recognition. The Frege point is not a thesis, or a conclusion derivable from premises, but an attainable insight; what is opposed to it is not a contrary arguable thesis, but merely one or other muddle that needs to be cleared up.

One well-known proponent of theories making out classes of statements to be truth-valueless was the late John Austin. He was excited by the discovery that many first-person present-tense statements serve to *enact* the performance which surface grammar might suggest they serve to *report*. (This was in fact a rediscovery: medieval discussion of sacramental forms of words, for example, had already made the point quite clearly.) 'I promise so-and-so', said in the right circumstances, is not an asserted proposition about the speaker, that he does promise so-and-so, but enacts the speaker's promise of so-and-so. Of course we must accept that there is this performative use of statements, and that it is non-normal: the same proposition does *not* occur asserted in the grammatical statement 'I promise so-and-so' and hypothesized in the protasis 'If I promise so-and-so'. But the notion of performative use assumed an exaggerated importance in Austin's mind, and got applied to cases that it clearly did not fit; moreover, as we shall see, even on his favourite topic of promising much that he said was confused or mistaken.

I shall approach the matter by first considering another verb with performative uses, the verb 'to assert'. The argument I put forward is modelled upon a *sophisma* of Jean Buridan. Suppose Johnson says 'I assert that the Earth is round'. Then we may argue that even though Johnson knows quite well that the Earth is round, all the same what Johnson asserts is false, is in fact a deliberate lie. For what Johnson *asserts that he is* asserting is: that the Earth is round. But what Johnson *is* asserting is *that he is asserting* that the Earth is round. Now plainly, that he is asserting that the Earth is round is one thing, and that the Earth is round is quite another thing: and Johnson must be quite aware of this. Moreover, the Earth's *being* round is not even logically implied by what Johnson *is* asserting, viz that *he asserts that* the Earth is round; and Johnson must know this too. So what Johnson asserts that he is asserting is different from, and not even logically implied by, what Johnson is asserting; so Johnson is knowingly making a false statement about what it is that he is asserting; so Johnson is telling a wanton lie.

Of course we ought not to find this argument demonstrative. In certain well-understood circumstances (we need not here spell them out in detail) such a statement in Johnson's mouth would certainly constitute an asserted proposition: not one saying anything about Johnson, but simply the proposition 'The Earth is round'; and similarly, if Johnson had said 'I assert that the Earth is flat', the asserted proposition would have been 'The Earth is flat'. The performative use of 'I assert' makes this difference, that Johnson is not in either case making an assertion as to what he is asserting; but it does not follow that Johnson was not asserting anything, on the contrary, he must have been asserting something about the figure of the Earth, truly or falsely as the case may be. All this is pretty obvious, and my elaboration of Buridan's *sophisma* may appear a tedious joke. Unhappily, real live philosophers have produced, with every appearance of seriousness, arguments no better than my Buridanic argument.

Let us now shift from Johnson to a rascally layabout Williamson, who goes round from farm to farm, telling each farmer 'I will kill your rats for £5, paid in advance'. Williamson has no knowledge of ratcatching and no intention of killing any rats; he simply spends the sums he receives, largely on drink. Plain men would certainly say Williamson lied in making his promise. I have however come across a philosopher who thought otherwise, and who even published his argument, to the following effect: Williamson could have made the same promise by saying not 'I will kill (etc.)' but 'I promise to kill (etc.)'. But if Williamson had used the latter form of words, he would not have been saying *falsely* that he promised to kill the rats, because he *would* have been promising that very thing, then and there. So if Williamson used the 'I promise' form, his utterance could not be a lie; and since the 'I will' form is a mere variant of the 'I promise' form, it too would give us a performative utterance not describable as truthful or mendacious. (I have altered the original example of a promise to my own; this does not affect the argument that was presented.)

This piece of reasoning is very much according to the mind of Austin: see *How to do Things with Words*, p. 11, for the denial of truth-value to promises in the 'I promise' form;[1] see also many passages in the later parts of that book where Austin—to my mind muddlingly—extends the notion of a performative utterance *from* cases where a verb is used to enact, rather than report, the kind of performance for which the verb stands, *to* cases where no such use of a verb is in question, but the same kind of performance is enacted. Surely Austin's influence is here visible.

Whatever the provenance of the argument, it is as fallacious as my Buridanic argument—only it was meant to win our assent, as mine was not. The argument:

> Williamson said 'I promise to kill the rats'; but Williamson did then and there promise to kill the rats; so Williamson was not lying

no more ought to produce any conviction than the argument:

> Johnson said 'I assert that the Earth is round'; but Johnson was *not* then and there asserting that the Earth is round (rather, asserting *that he asserted* that the Earth is round); so Johnson was lying.

Whether Johnson used the performative verb 'I assert' or merely said 'The Earth is round', he will have been asserting that the Earth is round, and his assertion is true just in case the Earth *is* round. Similarly, whether Williamson used 'I promise' or just 'I will', what he said was that he was going to kill the rats; since he had no intention, and indeed no skill, of ratcatching, this was a lie. Neither Johnson's nor Williamson's statement is made not to be an asserted proposition merely by the use of a performative verb. Since these are not, in my sense, normal statements, the proposition asserted is not a true or false assertion about what the speaker is then and there doing; but that nowise means it is not true or false at all. And if Williamson had made the same promise in the 'I will' form, there would be even less excuse for denying truth-value to what he said: in fact as little excuse as there would be for denying truth-value to Johnson's utterance if he had said simply 'The Earth is round'.

On one occasion when I was making these points, I was criticized for having confused 'I will', an expression of intention, with 'I shall', used to make plain future-tense statements. The criticism was relevant and successful only if it could show that 'I will' statements must in that case be denied to have truth-values. But how could this be shown? Why should expressions of intention lack truth-values? I cannot here discuss the nature of intention; it is at least not immediately obvious that we ought to reject the dictum of an English judge about intention—that the state of a man's mind is as much a fact as the state of his digestion. Anyhow, in fact I see no reason to doubt that 'I will' statements are normal statements about the future. The *shall/will* distinction appears to me largely an idiotism of idiom, and I welcome the fact that as a shibboleth marking out educated English speech the distinction seems to be sinking into disuse. Many languages use a future form of verbs that makes no such distinction. There are causal and epistemological differences between future events that are, and ones that are not, up to the choice of human individuals: but the futurity of events need not therefore be correspondingly different.

'I intend' statements, indeed, appear to me to resemble 'I promise' statements in being non-normal and in being what may be termed logical back-formations. Smith makes various 'I will' statements in circumstances in which others may appropriately say 'Smith intends . . .' or 'Smith promises . . .'; from this there develops the use by Smith, not of 'I will', but rather of 'I intend' to express an intention, or of 'I promise' to make a promise. This relation of 'I intend' said by Smith to 'Smith intends', or of 'I promise' said by Smith to 'Smith promises', is not the same as the relation of, for example, 'I smoke' said by Smith to 'Smith smokes', these last being (in my sense) normal statements. But there is a vital difference between 'I intend' and 'I promise'. To say 'I promise', in certain appropriate circumstances, just *is* to promise, and in these circumstances Smith's having said 'I promise' warrants others in saying 'Smith promises'. But to say 'I intend' *never* constitutes an intention; not even if the saying is done in the heart—a point that we must insist upon as against the sophistical Catholic moral theories of 'directing the intention'.

'I know', on which John Austin's curious view has become so celebrated, is in this respect like 'I intend', not like 'I promise': to say 'I know', with your lips or in your heart, can no more constitute your knowing than saying 'I intend' can generate an intention. Intention and knowledge are both states of a person, not episodic performances: whether they are ascribable to a person at a time has to be settled regardless of what is going on just then in his mind. Austin was disposed to assimilate 'I know' and 'I promise', but he was clearly wrong. Possibly what confused him was the fact that a man who claims to know and a man who is making a promise may alike say 'I give you my assurance on the matter': in the one case, assurance that such are the facts—in the other case, assurance that something will be done. To say such words *is* to give one's assurance; but, I repeat, to say 'I know' never is to know; so if this consideration influenced Austin, it did not give him good reason for his view.

Austin's account of 'I know' statements does not fit even the whole class of these. For it is characteristic of 'I know' to be followed not only by *that*-clauses but by indirect questions, and indeed by any sort of indirect question, theoretical or deliberative: of 'know' and indirect question, one may say they are intimately related; *sie sind Vetter*, in Wittgenstein's phrase. Indeed, 'x knows that p' may be glossed as 'Both (x knows whether *p*) and *p*'; so it is easy to explain the construction of 'knows' with *that* clauses in terms of its construction with an indirect question. The converse type of explanation is not at all easy to work; it is fairly well known, for example, what difficulty Hintikka and others have encountered over analysis of knowing *who* did in terms of knowing *that so-and-so* did. (I owe the considerations in this paragraph to unpublished work by Christopher Hookway, of the University of Birmingham.)

Now if the schoolmaster says 'I know that Jones minor broke the window', it may be said perhaps that he is telling his audience that Jones minor broke the window and staking his authority upon the truth of this. But if he says 'I know who broke the window', he is not even telling anybody who it was that broke the window, still less staking his authority on this. So here we have a large class of 'I know' statements for which, manifestly, nothing like Austin's account is going to work. We may notice that there is no corresponding construction of 'I promise' with indirect questions: *'I promise who shall get the prize' (or *' . . . is going to get the prize') is barely intelligible, and even if it were said in the appropriate circumstances by a person who was in a position to give a promise as to who should get the prize, this 'I promise' statement could not constitute the giving of such a promise.

A further point, which I have made elsewhere, is that an allegedly truth-valueless performative utterance may easily go together with other premises to yield a conclusion in a way that suggests that all the premises are after all normal statements. In his book on Frege, Michael Dummett chivalrously defends his Oxford colleagues against my strictures on this point: he elaborately argues that even if a statement '*p*' is not normal and utterance of it does not give us an asserted proposition at all, we can even so explain the occurrence of an equiform sentence in 'If *p* then *q*' and the apparent detachability of the antecedent as in *modus ponens*.[2]

Dummett's defence of the Oxford views would work, perhaps, for the example I gave in *Logic Matters*, p. 268;[3] but the defence is effective only on a very narrow front, and can easily be bypassed. The following simple modification of that argument already bypasses Dummett's defence:

I am no art expert, and I know Smith's Vermeer is a fake.

If somebody who is no art expert knows Smith's Vermeer is a fake, then Smith's Vermeer is a clumsy forgery. *Ergo*, Smith's Vermeer is a clumsy forgery.

Only by a highly contrived and quite unconvincing analysis could the validity of this inference be explained so as to avoid treating the statement:

I know Smith's Vermeer is a fake.

as a normal statement, true if and only if the speaker knows Smith's Vermeer is a fake. With this I leave this *genre* of philosophizing: I could have subjected many other examples of it to similar destructive analysis. Such theories, as I have said before, are plausible only for thinkers who suffer from deficiency diseases of thought, because they feed their minds on an unbalanced diet of examples.

I now turn to the theory of logical islands (of course not *one* theory, but once again a *genre* of theories). It is held that statements in general divide exclusively and exhaustively into two kinds, *A* and *B*. Each kind of statement is supposedly closed under negation; and supposedly no inference of *A* statements from premises all of which are *B* statements, or *vice versa*, is possible. This type of theory, as I said, admits of positive logical refutation: at least it does so if a disjunction can be formed one of whose two members is equiform, and has the same content, with an *A* statement, and the other with a *B* statement. Let the first member be abbreviated to '*p*' and the second to '*q*': is '*p* or *q*' an *A* statement or a *B* statement? If it is an *A* statement, then in view of the valid inference '*q*, *ergo p* or *q*' an *A* statement will have turned out inferable from a *B* statement. Suppose then that '*p* or *q*' is a *B* statement. Since the class of *B* statements is closed under negation, 'not *q*' will be a *B* statement just as '*q*' is. But then the valid inference '*p* or *q*, not *q*, *ergo p*' will lead us from two *B* premises to an *A* conclusion. Either way the rule fails: it is logically impossible to rule out the inference of *A* conclusions from *B* premises.

The only way to escape this conclusion is to hold that *A* and *B* statements are so much isolated from each other that no disjunction of these pairs is well-formed. This would be a desperately implausible thesis, except indeed on one kind of view falling into the *genre* we are discussing: namely that *B* statements are purely indicative, whereas *A* statements, in spite of their grammatically indicative form, either are disguised imperatives or at least entail imperatives. For a disjunction of an *A* statement and a *B* statement must in that case either be a disguise for or entail a disjunction between an indicative and an imperative; and such a disjunction is on the face of it ill-formed; so on this view *A-B* disjunctions could not be sensibly formed either.

Now why is a disjunction of an indicative and an imperative ill-formed, whereas a disjunction of two indicatives or two imperatives is well-formed? The question rests, I think, upon a misapprehension: assertoric or imperative force never attaches logically to one or other clause within a complex sentence, but solely to the whole period. Grammar disguises this from us, but it is easy to penetrate the disguise. Regardless of the indicative or imperative mood of the clauses, we can see that all of these are equivalent:

(a) Either do not walk on the grass or wear your Wellingtons.
(b) If you walk on the grass, wear your Wellingtons.
(c) Do not walk on the grass unless you wear your Wellingtons.

The imperative force attaches not to the separate clauses but only to the whole: this may be brought out by using Hare's technique of separating the 'phrastic' and the 'neustic':

Your not walking on the grass or alternatively your wearing your Wellingtons—please!

and then the equivalence between (a), (b), and (c) is to be seen as simply the equivalence between 'either not *p* or *q*', 'if *p* then *q*', and 'not *p* unless *q*'; the difference between main and subordinate clauses, and the grammatical rule whereby the imperative mood attaches only to the verb of a main clause in a sentence that has as a whole imperative force, are of no importance for logic. Hare must be given credit for expounding this analysis, parallel to Frege's doctrine of the assertion sign, in *The Language of Morals*; but the exposition was marred by some confusing passages where Hare tried to explain how imperative force can attach to the antecedent and consequent of a hypothetical,[4] which on my view is as much excluded by a correct analysis as would be the attachment of assertoric force to such clauses.

It may be objected that this sort of equivalence will not work for many hypotheticals:

If it rains, wear your mackintosh

clearly cannot be transformed, by analogy to the above example, into either of these:

*Either let it not rain or wear your mackintosh

*Let it not rain unless you wear your mackintosh.

But the explanation is easy. The Creator in *Genesis*, and Christ in the Gospels, are represented as simply commanding that things shall happen thus and so; conjurors' patter makes believe to do the like, and absolute tyrants may utter such a form of command seriously. (Emperor Christophe of Haiti would simply command that some hole in a road be filled up; he did not bother to address his command *to* anybody; the sanction was that if the hole were not filled up, a few randomly chosen people in the district would be shot.) Normally, however, a command is addressed to somebody who is supposed to bring a state of affairs about, and is not a naked decree that the state of affairs shall be; so an adequate logic of imperatives would have to use a theory of the verb 'to bring it about that'. When we see this, the puzzle disappears. For

Bring it about that if it rains you are wearing your mackintosh

is indeed logically equivalent to

Bring it about that either it does not rain or you are wearing your mackintosh

and if the latter is addressed to an ordinary man who has no control over its raining or not raining, then the only way it can be fulfilled or unfulfilled through his action is by his wearing or not wearing his mackintosh.

We need not doubt, then, that imperative force attaches only to complete utterances, not as grammar suggests to clauses; and this is good reason for denying that a disjunction can be well-formed whose disjuncts are one imperative and the other indicative (logically, that is, rather than grammatically). But we could not extend this to clauses whose verb is 'ought' or the like: there appears no reason why these should be barred from being one side of a well-formed disjunction whose other disjunct is a factual statement, for example:

Either Smith's wife is deceiving him, or Smith ought to consult a psychiatrist.

Indeed, in sharp contrast to imperative force, which can attach only to complete periods, 'ought' can figure as an operator in mere sentence-fragments, which could not be turned into freestanding sentences:

> If the chairman allows committee members to talk longer *than they ought*, then the committee will not finish the agenda that *this meeting ought to dispose of.*

Hare has often claimed that 'ought' statements, or at least a subclass of them which he would count as expressing 'synthetic evaluative judgments', entail imperatives: if he were right, then indeed a disjunction between an 'ought' statement and a factual statement would be as ill-formed as one between an imperative and an indicative, and the backbone of my argument against logical islands would be broken. But Hare has never, to my mind, satisfactorily answered the criticisms of his views on the logic of imperatives; and he has not even sketched a mixed deontic and imperative logic that would show how deontic utterances entail imperatives, according to what rules of inference. In these circumstances his thesis wholly lacks rational support. He seems to think the thesis can be upheld through making it part of the *definition* of 'synthetic evaluative judgments' that they do entail imperatives. This has of course a great advantage over the formulation of a logical theory: in fact the advantage Russell ascribes to postulation (see *Introduction to Mathematical Philosophy*, p. 71).[5]

Hare has indeed noticed, and tried to explain away, the occurrences of 'ought' in sentence-fragments, mentioned above, which on the face of it stand irremovably in the way of assimilating 'ought' to a sign of imperative force. Discussing the example:

> At the very moment when he ought to have been arriving at the play, he was grovelling underneath his car five miles away,

he sketches the following approximate analysis:

> At the very moment when most people (myself included) would agree in saying 'He *ought* to have been arriving at the play', he was grovelling under his car five miles away.[6]

This is, as the medievals would say, *extorta expositio*. Incidentally, the italicized 'ought' is meant to indicate not emphasis but a technicality of Hare's—an artificial word in some way similar in its use to the natural word 'ought'. This adds to the implausibility of the analysis, but I need not dwell on that. The whole idea of the analysis is clearly wrong: we need only spell out a corresponding analysis of the example about the committee:

> If the chairman allows more than one committee member to talk so long that most people (including myself) would agree in saying 'He *ought* to shut up', then certain agenda about which most people (including myself) would agree in saying 'They *ought* to be disposed of at the meeting' will never be finished at the meeting.

I turn from this unsuccessful attempt to divide discourse into logical islands to another attempt: the attempted insulation of *A* statements, which are either necessary or impossible, and *B* statements, which are contingent. For this example, fortunately, there is no need to go into tricky arguments about bridging the gulf with a mixed disjunction. There are much simpler ways of showing that necessary *A* statements can be inferred from contingent *B* statements. Let '*p*' be short for any contingent statement; from '*p*' follows '*p* or *q*', whatever '*q*' may be short for; if we identify the interpretations of '*q*' and 'not *p*', we have a necessary statement, a case of the Law of Excluded Middle, logically following from a contingent one. Or again, however contingent 'All ruminants are cloven-hoofed animals' and 'All cloven-hoofed animals are ruminants' no doubt are, they yield by syllogism the necessary statement 'All ruminants are ruminants'. It would take a quite extraordinary non-standard logic to debar these and innumerable other kinds of inference which lead from contingent premises to necessary conclusions.

I cannot certainly account for the prevalence of this particular error. I suspect it may be a remote hangover of a vaguely remembered rule in Aristotle's modal logic: that you cannot get a 'necessary' conclusion except from a 'necessary' premise. The word 'necessary' in quotes here is a rendering of Aristotle's '*anankaia*'; in this use, a proposition's being 'necessary' means not that it *is* a necessary truth, but only that it *says* something is necessarily so. 'Some philosophers are necessarily donkeys' would in this sense be 'necessary', although it is false; and 'All men are men' would not be 'necessary' in this sense, although it is necessarily true. I need not discuss the rule of modal syllogistic; only by muddled equivocation could its soundness, if established, be held to justify the belief that a necessary conclusion cannot logically follow from one or more contingent premises.

Another likely source of the error is a misreading of the doctrine of truth-functions, particularly as presented in Wittgenstein's *Tractatus*. Here I need not resort to conjecture: I have myself heard a philosopher read an elaborate paper on the *Tractatus* in which the key of the whole interpretation was the logical insulation of tautologies and contradictions on the one hand and contingent statements on the other; he not only believed that this insulation was what Wittgenstein maintained in the *Tractatus*, but was actually prepared to maintain it in his own person. Fortunately I need not discuss the historical question, though for the record I must say that the grounds for ascribing the insulation doctrine to Wittgenstein seem exceedingly shaky. In any event, only by confusion could anybody suppose that the doctrine of truth-functional compounds requires such insulation: it requires the very opposite, and therefore could not also require the insulation doctrine unless it were internally inconsistent, which nobody is going to show. As I said, the tautology '*p* or not *p*' logically follows from '*p*', however contingent '*p*' may be; and this is simply a part of truth-functional logic.

I had intended when I began writing this paper to document the influence on philosophers of our time of the insulation doctrine as applied to necessary-or-impossible versus contingent statements. One very surprising example is to be found in Braithwaite's *Scientific Explanation*,[7] p. 294. But my paper is already long enough. I end by restating the conclusion of the whole matter, already anticipated.

There is a great over-extension in recent philosophy of the idea that grammatical statements need not have truth-value; many particular applications of this idea are misguided. There is no foundation at all for the idea that there are logically insulated kinds of truth: truth is just truth. For a Christian, truth is God's truth; and for unbelievers too, truth is often found worth living and dying for. There was a wartime slogan in Poland that I heard of: *walczymy za Prawdę i Polskę*, we fight for Truth and Poland: a slogan of which those who upheld logic and other learning at such peril in the underground Universities showed themselves worthy. A lady who wished to insulate theological truth from secular truth once asked whether one would or should give one's life for a truth about Banach spaces. Well, there are barbarous regions in the United States where people believe that $\pi = 3$ because they think that by implication 'it says so in the Bible' (with reference to the dimensions of King Solomon's bath); and there a man might have to suffer at least as a confessor, if not as a martyr, for saying that $\pi \neq 3$. As for what I would do in such a situation, I can only borrow words that a worldly cleric of Tudor times is said to have used, being asked to put himself in Joseph's shoes when tempted by Potiphar's wife: Marry, I know not what I *would* do, but I know well what I *should* do.

Notes

1 J. L. Austin, *How to do Things with Words* (OUP, Oxford, 1975).
2 M. Dummett, *Frege: Philosophy of Language* (Duckworth, London, 1973), pp. 327–354.
3 P. T. Geach, *Logic Matters* (Basil Blackwell, Oxford, 1973).
4 R. M. Hare, *The Language of Morals* (OUP, Oxford, 1964), p. 34f.
5 B. Russell, *Introduction to Mathematical Philosophy* (Allen & Unwin, London, 1919).
6 R. M. Hare, *op. cit.*, p. 193ff.
7 R. B. Braithwaite, *Scientific Explanation* (CUP, Cambridge, 1953).

[15] TIME, TRUTH AND NECESSITY

G. H. von Wright

Part I

Few single chapters in the writings of philosophers have challenged so much discussion as the ninth chapter in Aristotle's Περὶ ἑρμηνείας. This is due, no doubt, partly to the intrinsic interest of the problems and partly to difficulties in understanding the text. A discussion of the chapter may be a contribution either to the discussion of the philosophic problems or to the interpretation of Aristotle. Or it can be a contribution to both—as is the case with Elizabeth Anscombe's influential paper 'Aristotle and the sea battle' in *Mind* (1956).

I cannot myself claim authority in matters of Aristotelian scholarship. Let me say, unpretentiously, that I think I understand what is intriguing the philosopher in the celebrated chapter in question and how he solves the puzzle. I find Aristotle here, as always, exceedingly clearheaded and there is nothing eccentric or strained about the solutions he offers. Aristotle's genius is that of a *sober* mind. He is never cryptic or enigmatic, as Plato or Wittgenstein may be said to be. Therefore, one may not find him as 'exciting' as one finds them. But people who think Aristotle pedestrian or even philistine seem to me simply to suffer from a blind spot in their appreciation of intellectual greatness.

Part II

Hintikka has argued[1] that the problem which has traditionally been associated with the ninth chapter of *De Interpretatione* is not the problem which primarily worried Aristotle. In what Hintikka calls 'the traditional view', the problem is about the validity of the Law of Excluded Middle for contingent propositions about future events. Hintikka dubs this 'the problem of future truth'. It can be paraphrased as follows.[2]

Let it be granted that it is true today that there will be a sea battle tomorrow or there will not be a sea battle tomorrow. Is it not then also either true today that there will be a sea battle tomorrow or true today that there will not be a sea battle tomorrow? That is: does it not follow then that one of the two disjoined propositions—we may, however, not yet know which one—is already true today? And, if so, are we not committed to accepting determinism? Because does what has been said not mean that that which is going to be is already before it happens 'settled', predetermined, unavoidably going to be?

Propositions such as that there will be a sea fight or an earthquake tomorrow, or at some future date *t*, are, one would think, contingently true or false. But from assuming the unrestricted validity of what has become known as the Law of Excluded Middle it seems to follow that such propositions as those mentioned are, if true, necessarily true and, if false, impossible. This strikes one as absurd. Something must have gone wrong in the progression from the initial assumption to the deterministic conclusion.

This description of the problem makes only oblique and peripheral reference to certain notions which are prominent in Aristotle's discussion in the famous ninth chapter. I am thinking chiefly of the notions of necessity and of omni-temporality and of time generally. Their rôle will be discussed presently. It seems to me, however, that it was precisely the problem as described above which puzzled Aristotle—as it has confused and worried many logicians and philosophers since. One of them was the Pole Jan Łukasiewicz.

Part III

On Łukasiewicz's diagnosis of the situation,[3] what is wrong with the chain of thought in our above presentation of the problem is the step from the assumption that the disjunction of a given proposition and its contradictory (negation) is true to the conclusion that either the proposition in question is true or its contradictory is true, i.e., the proposition itself false. The assumption, in its general form, is the Law of Excluded Middle. It can be accepted. For the purported conclusion Łukasiewicz coined the name Principle of Bivalence. If by the falsehood of a proposition we understand the truth of its contradictory (negation), then the principle, in its general form, says that any given proposition has one of two truth-values, is either true or false. This principle, Łukasiewicz thought, is not universally valid. It does, for example, not hold good for contingent propositions about the future. (Łukasiewicz also expressed this by saying that such propositions and their negations do not stand in a contradictory relationship to one another.)

Łukasiewicz's rejection of the Principle of Bivalence was a starting point for his grand conception of a many-valued logic, related to 'classical' two-valued logic in a way analogous to the relation between non-Euclidean and Euclidean geometry. Lukasiewicz was not the first to entertain the idea of a polyvalent logic. But he gave the decisive impetus to its modern development.

Łukasiewicz was also of the opinion that he was following Aristotle in accepting the one but rejecting the other of the above two principles. The view that Aristotle's own 'way out' consisted in a denial of the Principle of Bivalence has a long tradition and supporters also among modern scholars. It is sometimes referred to as the 'orthodox'[4] or the 'traditional'[5] interpretation of Aristotle's text. It is contested by Anscombe, and also by Hintikka, and I think they are absolutely right in this.

On my reading of the text, there is actually only one passage which seemingly supports the view (of *De Int.* 9) of Łukasiewicz. It occurs at the end of the chapter. It has always baffled me, since it seems to be at odds with what Aristotle has already clearly stated as his position. In 19a36–38 we read: *τούτων γὰρ ἀνάγκη μὲν θάτερον μόριον τῆς ἀντιφάσεως ἀληθὲς εἶναι ἢ φεῦδος, οὐ μέντοι τόδε ἢ τόδε ἀλλ᾽ ὁπότερ᾽ ἔτυχε καὶ μᾶλλον μὲν ἀληθῆ τὴν ἑτέταν, οὐ μέντοι ἤδη ἀληθῆ ἢ φευδῆ.* Or, in Elizabeth Anscombe's translation:[6] 'For such things it is necessary that a side of the antiphasis should be true or false, but not this one or that one, but whichever happens; and that one should be true rather than the other; but that does not mean that it is true, or false.'

The first sentence in the quoted passage clearly sounds like an affirmation of the Law of Excluded Middle (for contingent things, *ἐπὶ τοῖς μὴ ἀεὶ οὖσιν ἢ μὴ ἀεὶ μὴ οὖσιν*). Aristotle, moreover, says that the disjunction under consideration is necessarily true. But in the last sentence he seems to be denying the Principle of Bivalence. The impression, I should say, is strengthened by Miss Anscombe's translation: 'that does not mean that it is true, or false'. She omits the word ἤδη or, rather, conceals it in the phrase 'that does not mean'. The most obvious translation would be 'already' or 'yet'. We should then understand Aristotle as saying that, although the disjunction is necessarily true, the disjuncts taken by themselves are 'not yet' true or false. As I shall argue below (Part VI), this 'temporalization' of the truth-values makes good sense and removes the impression of conflict with the Principle of Bivalence. Anscombe, however, firmly disputes this translation, but unfortunately without convincing arguments.[7]

Part IV

As has often been noted, the distinction between the two principles is of no avail in dealing with the problem posed by future contingencies. When applied to the notion which I propose to call *plain truth*, the phrase 'it is true that' is semantically otiose when prefixed to a declarative sentence. 'It is true that *p*' and '*p*' say the same. Therefore 'it is true that there will be a sea battle tomorrow or there will not be a sea battle tomorrow' and 'either it is true that there will be a sea battle tomorrow or it is true that there will not be a sea battle tomorrow' both say exactly the same, viz. that there will be a sea battle tomorrow or there will not be a sea battle tomorrow. Generally speaking: for the notion of 'plain truth' the Law of Excluded Middle and the Principle of Bivalence coincide. If the 'orthodox interpretation' is to drive a wedge between the two principles, then we may safely conclude that it leaves the problem which worried Aristotle unsolved and also, I think, that Aristotle himself did not wish his position to be thus interpreted.

Let us therefore look for a better solution to Aristotle's problem. In order to find it we must relate the notion of truth to two other notions which also figure prominently in the text. These, as already indicated, are the notions of time and necessity.

Part V

We sometimes make a *tensed* use of the phrase 'it is true that'. We say: 'it is now (sometimes, always) true that—', 'it will be true that—', 'it was true that—'. When the notion of truth involved is 'plain truth' I shall call such tensed use *spurious*. By this I mean the following: To say that it is now (sometimes, always) true that it is raining means that it is true that it is now (sometimes, always) raining which means that it is now (sometimes, always) raining. To say that it will be true that a sea battle takes place is to say that it is true that a sea battle will take place (in future) and this again is to say, simply, that a sea battle will take place. By calling the tensed use 'spurious' I mean that it can always be replaced by the form of words 'it is true that' which can then be omitted as otiose.[8]

Generally speaking: 'it is true at *t* that *p*' means the same as 'it is true that *p* at *t*' and both mean, simply, that *p* at *t*. Here '*p*' stands for an 'open' declarative sentence such as 'it is raining' or 'Socrates is sitting' which does not, as it stands, express a true or false proposition but does this only when coupled with an 'absolute' location *t* in time, for example, a historical date.[9] Open sentences (propositions) are often said to have a variable truth-value, or to be true on some occasions and false on others. This is an easily misleading *façon de parler*.

To call a tensed use of the phrase 'it is true that' spurious is to say that this phrase is *atemporal*. ('Truth is eternal.') *A*temporality must not, however, be confused with *omni*-temporality. It is therefore misleading to say that, if it is true that *p* at *t*, then it is *now and always* true that *p* at *t*. To say that it is true at *t′* that *p* at *t* is not to say anything more than that it is true that *p* at *t*, i.e., to say that *p* at *t*. It follows that it is in the same way misleading to say that, if (it is true that) there will be a sea battle tomorrow, then it was true already 10,000 years ago that there was going to be a sea battle on that day. But it is of course true and in no way misleading to say that, if there is going to be a sea battle tomorrow, then a man who 10,000 years ago had predicted or said that there was going to be a sea battle on that future day would have been *right*, i.e., would have spoken the truth.

The notions of omnitemporality and necessity are closely related in Aristotle; he might even have thought that if something is omnitemporally true then it is necessarily true (see Part VIII). Did Aristotle confuse atemporality with omnitemporality, or perhaps even confuse the atemporal character of ('plain') truth with necessary truth? I do not think he can be accused of having done this. But the discussion in the ninth chapter is, I think, evidence that he was puzzled and struggling against a confusion of which some others (perhaps the Megarians) actually were guilty.

Part VI

Once one sees clearly what the atemporal character of truth amounts to and that the atemporal must be distinguished from the omnitemporal, there should no longer exist a temptation to draw deterministic conclusions from the admitted or assumed unrestricted validity of the Law of Excluded Middle. Still—a feeling of puzzlement may remain. For me—and perhaps for Aristotle too—this is connected with the fact that the phrase 'it is true that' also has a temporalized use which is *not* spurious. This use is involved, for example, when we say with emphasis 'it is *already* (now) true that—' or 'it was already (then) true that—'. This can be merely a misleading way of emphasizing that something is *true.* Then the notion of truth involved is 'plain truth'—and 'already true' means 'true'. But the phrase may also express something which can be paraphrased by saying that it is (was) already 'fixed' or 'settled' or 'certain' or 'necessary' that such and such is (was) going to be true. Truth, thus 'fixed', is not 'plain truth' but something more complex and stronger.

When 'true' is used in this other way then it *is* already (today) true that there will be a sea battle tomorrow or there will not be a sea battle tomorrow, but it may nevertheless be neither already true that there will be a sea battle tomorrow nor already true that there will not be a sea battle tomorrow. Thus, for this other notion of truth, the Law of Excluded Middle holds good, whereas the Principle of Bivalence is not universally valid. But this is but another, and easily misleading, way of saying that the necessary ('fixed', 'settled') truth of the disjunction of two propositions is compatible with the contingent ('not-yet-settled') truth or falsehood of the disjuncts individually. And this is, almost in so many words, affirmed in 19a27–33 by Aristotle himself for the special case of an antiphasis:

> For it is necessary that everything should be or not, and should be going to be or not. But it is not the case, separately speaking, that either of the sides is necessary. I mean, e.g., that it is necessary that there will be a sea battle tomorrow or not, but that it is not necessary that there should be a sea battle tomorrow, nor that it should not happen.

I submit for consideration that when the text a few lines later inserts the word *ἤδη* before *ἀληθῆ ἢ ψευδῆ*, this happens in order to indicate that we are not now concerned with the 'plain truth' of the two sides of the antiphasis, but with their predetermined or necessary truth.

Part VII

In 19a 23–24 the text reads *Τὸ μὲν οὖν εἶναι τὸ ὂν ὅταν ᾖ, καὶ τὸ μὴ ὂν μὴ εἶναι ὅταν μὴ ᾖ, ἀνάγκη*. In Ascombe's translation: 'The existence of what is when it is, and the non-existence of what isn't when it isn't is necessary'. The same necessity also belongs to everything which has been, that is, to everything past and true. There can be no doubt that Aristotle would have subscribed to the first 'premiss' in the famous Master Argument of Diodorus: *Πᾶν παρεληλυθὸς ἀληθὲς ἀναγκαῖον*. The same idea is also echoed in the schoolmen's 'quod fuit non potest non fuisse' and 'omne quod est, quando est, necesse est esse'.

Elizabeth Anscombe notes (p. 12) that this way of speaking about necessity 'has a sense which is unfamiliar to us'. Perhaps this is so; at least it is a sense which is hardly familiar to modern modal logicians. But it is not a sense which is contrary to a common way of understanding and even of speaking about things. So what does it mean?

On this question I disagree with Anscombe. She writes (p. 7):

> A modern gloss, which Aristotle could not object to, and without which it is not possible for a modern person to understand his argument, is: and cannot be shown to be otherwise. It will by now have become very clear to a reader that the implications of 'necessary' in this passage are not what he is used to.

I dare say that Aristotle would not have objected to her 'modern gloss'. But I do not think that, without this twist, 'it is not possible for a modern person to understand his argument'. The argument certainly has the epistemic *implications* which Anscombe attributes to it. Something important, however, gets lost if we try to understand the idea of the necessity of that which is or has been, *qua* present or past fact, exclusively or primarily in *epistemic* terms. In order not to trivialize the idea we must relate it to notions of potentiality, 'real' possibility, and—if we wish to be 'modern'—also to ideas about causation.

That everything which has actualized is also necessary should, I think, be understood to mean that once it is there it is 'fixed and settled'. *Nothing that happens subsequently can make a difference to it*. This idea is reflected in such common sayings as 'facts cannot be changed' or 'what is done cannot be undone' (which is true or false depending upon how we understand it).[10]

This is not as trivial as it may sound, because if causation could work backwards then something which is a fact at a certain time *t* may later cease to be a fact *at that same time t*. If backward or retroactive causation is possible, then things which were not the case could be caused to have been the case by something which happens later; and something, which was, could be caused not to have been. This sounds extremely odd. It may be thought to be 'contrary to reason'. Perhaps it is. But the question is serious and has been much debated.

One can therefore understand the idea that past and present truths are necessary as a way of denying retroactive causation—and of affirming, obliquely, that causation works in the direction of time. ('First the cause, then the effect.') The past, one could also say, is a *closed* linear order of successive states of the world, up to and including the present. The future, it seems, is different. It may be *open*, have room for alternative developments, not yet 'fixed and settled'.

Epistemically, the past is just as 'contingent', 'open to alternatives', as the future. We may not know whether there was a battle at Salamis then, or not—just as we do not know for certain whether there will be one in future. But if a battle took place then, nothing we can do nor anything else that may happen can change this *fait accompli*. In this sense it is necessary.

I shall refer to the necessity, thus understood, of past and present truths as the necessary character of *truth as fact*.

Potentiality is forward-looking, 'for the future'. My coat may be cut in halves or it may not be cut but wear out first. This is Aristotle's famous example. Did he mean to say *only* that we do not know what will be the fate of my coat ('it remains to be seen')—or did he wish to say that the two possibilities, as it were, 'reside' in the nature of things here? If he had been a determinist he would have said that the contingent nature of future events is only epistemic. In fact everything that happens comes about of necessity. The fate of my coat is already settled. This, however, Aristotle did not say, nor did he wish to say it.

Part VIII

A few lines after having stated the necessary character of truth as fact Aristotle adds (19a26–27): 'For it isn't the same: for everything that is to be of necessity when it is, and: for it simply to be of necessity'. Aristotle evidently thought that by observing a distinction between εἶναι ἐξ ἀνάγκης ότε ἐστί and ἁπλῶς εἶναι ἐξ ἀγάγκης one could avoid a commitment to determinism. But how shall we understand this ἁπλῶς εἶναι ἐξ ἀγάγκης?

Perhaps a better rendering of ἁπλῶς here would be 'without qualification', where the qualification in question is temporal.[11] Because the contrast which Aristotle is stressing is between ὅτε ἐστί and απλῶς, that is between something being necessary 'when it is' and being necessary without this qualification, 'simply'. This contrast is, in turn, related to the distinction between an open and a closed sentence such as 'it is raining' on the one hand, and 'it is raining at Cambridge on 28 February 1978' on the other hand (see Part V).

It is obvious, and has been noted before, that when Aristotle discusses things which may be true or false he often has in mind temporally unqualified sentences (propositions). Only when one takes note of this can one make sense of such locutions of his as that sometimes 'the affirmation is not true rather than the negation; and with other things one is true rather and for the most part' (19a20–24). That one of a pair of contradictory propositons is true 'rather and for the most part' must mean that, on an unspecified number of repeated occasions, one member of the pair, either the kataphasis or the apophasis, turns out true more often than the other. (Aristotle's 'frequency theory' of probability.)

Let '*p*' stand for an open sentence such that '*p* at *t*', but not '*p*' by itself, says something true or false. Suppose further that, for some value(s) of *t*, it is true that *p* at *t* and, for some other value(s) of *t*, it is false that *p* at *t*. This would be sufficient grounds for pronouncing contingent 'without qualification' (ἁπλῶς, *simpliciter*) the proposition that *p*. If, therefore, a proposition is necessary without qualification, it must *not* happen that it sometimes turns out to be true and sometimes false when qualified in time, but it must always turn out to be true, be true omnitemporally.

If the proposition, for example, that it is raining were necessary ἁπλῶς then it would keep on raining 'from eternity to eternity'. Aristotle would presumably have said of some natural processes, for example, the movements of the heavenly bodies, that they are in this sense 'necessary'. They immutably go on for ever.

Aristotle, moreover, also seems to have inclined to the opinion that if something is omnitemporally true, then it is necessarily true. This is somewhat controversial, however, and I shall not go into the controversy here. But the following observation may be worth making: Since that which is necessary ἁπλῶς is omnitemporal and since everything which is is necessary when it is, it follows logically that, if something is omnitemporally true then it is always necessary. And this is almost but not quite the same as saying that omnitemporality is tantamount to necessity. In order to have a full equivalence here we should still have to show that if something is always necessary then it necessarily always is, i.e., that omnitemporal necessity entails necessary omnitemporality. Whether one can establish this identity on Aristotelian 'premisses' is not clear to me.[12]

Be this as it may, Aristotle is firm about the existence of things which are *not* necessary ἁπλῶς and therefore, in this sense, contingent. From the common sense point of view he is, of course, right. It simply *is not* raining perpetually but only 'from time to time'. Perhaps Aristotle thought that the distinction between the necessity of *truth as fact* and necessity *simpliciter*, in combination with the observation that certain things are not always but only sometimes, was all that was needed to refute determinism. If this was how he thought he was mistaken, however.

Part IX

The idea of determinism can be explicated in more than one way. It is common to describe determinism as the view that nothing happens without a cause or, which is the same, that everything which happens has a cause. 'Cause' is then usually understood to mean an efficient (and not a final) cause. This explication of determinism is not the best one when discussing topics in ancient or mediaeval philosophy. It is too closely tied to notions of causation and *causal laws* which, I think, are on the whole alien to thinking before the Renaissance.

It is more useful for our purposes to regard determinism as a view to the effect that everything which is or happens is *predetermined*. This should then be understood as follows: it is predetermined that p at t if, and only if, some time before t it is 'already true' that p at t. Instead of 'already true' we can also say 'fixed' or 'settled' or 'certain' (when the word is stripped of epistemic connotations) or 'necessary' (see Part VI). I shall here, in agreement with common practice among philosophers, opt for the term 'necessary'.

One can further distinguish between a weaker and a stronger form of determinism *qua* predetermination. The weaker version says that it is predetermined that p at t if, and only if, *there is* a time t' earlier than t such that it is necessary at t' that (it will be true) that p at t. The stronger version says that it is predetermined that p at t if, and only if, at *any* time t' earlier than t it is necessary that p at t. This distinction is important but I do not think that it is relevant to a discussion of the ninth chapter of *De Interpretatione*.

The definition of determinism as predetermination is wider than the definition in terms of efficient causes. *We* tend to think that if it becomes necessary at t' that p at t, then this is due to the happening at t' of something which *causes* that p at t. For example, that q at t' causes that p at t. Someone might wish to add: 'by virtue of a (causal) law'. The existence of causes and causal laws is surely a *form* of predetermination. Does predetermination *require* causes (and laws)? The question is worth discussing but may be left open here.

The idea that everything which is, when it is necessary, does not mean that truth as fact is predetermined. Aristotle certainly did not wish to say that everything past or present is or has come to be of *necessity*. On the contrary, he denies this. *φανερὸν ἄρα ὅτι οὐχ ἅπαντα ἐξ ἀνάγκης οὔτ' ἔστιν οὔτε γίγνεται, ἀλλὰ τα μὲν ὁπότερ' ἔτυχε* (19a 18–19); 'it is clear that not everything is or comes about of necessity, but with some things "whichever happens". (The Loeb translation renders *ἀλλὰ τὰ μὲν ὁπότερ' ἔτυχε* by 'Cases there are of contingency'.)

Aristotle distinguishes between something being *necessary*, *ἀναγκαῖον* and something being or happening *of necessity*, *ἐξ ἀνάγκης*. That everything which was is, so to say 'in retrospect', necessary does not mean that it came to be of necessity. It is not quite clear (to me) how the Aristotelian phrase *γίγνεται ἐξ ἀνάγκης* ought to be understood. But I see no reason to doubt that it means what I have here called 'predetermined' of 'antecedently necessary'.

One may now accordingly distinguish two kinds of contingency. If the proposition that p is neither necessary nor impossible 'without qualification' or *ἁπλῶς*, then it is contingent. If again neither the proposition that p at t nor its negation, the proposition that not-p at t, is predetermined, true *ἐξ ἀνάγκης*, then the proposition (and its negation) is contingent.

Now assume that it is true that p at t_1 but false that p at t_2, t_1 and t_2 being two arbitrarily selected locations in time. Then the proposition that p is neither necessary nor impossible *ἁπλῶς*, but contingent. But this fact is fully compatible with the possibility that it is necessary at t'_1 that p at t_1 and necessary at t'_2 that not-p at t_2, where t'_1 is a time before t_1 and t'_2 a time before t_2. Generally speaking: the contingent nature of the proposition that p is fully compatible with the predetermined truth or falsehood of the proposition that p at t for every value of t.

That there is *change* in the world means that there are propositions that p such that for some values of t it is true and for some other values false that p at t. If one accepts as a fact that change occurs, then one can also be sure that there is contingency in the sense that not everything is either necessary or impossible *simpliciter*, ἁπλῶς. But this is no proof that there is contingency in the sense required for refuting determinism, that is, no proof that everything does not, after all, happen ἐξ ἀνάγκης.

I do not know whether Aristotle can be said to have proved that change occurs, but he certainly did not wish to deny it. (As some other philosophers, for reasons of conceptual puzzlement, had done). By assuming change Aristotle can be said to have 'saved contingencies'. Aristotle also *denies* determinism. But nothing which is said in the Ninth chapter about saving contingencies amounts to a *refutation* of determinism.

Notes

1 J. Hintikka 'The once and future sea-fight: Aristotle's discussion of future contingents in *De Interpretatione* IX', *The Philosophical Review* 73, (1964); reprinted with minor changes in J. Hintikka, *Time and Necessity, Studies in Aristotle's Theory of Modality* (Clarendon Press, Oxford, 1973).
2 In this paraphrase I follow my paper 'Determinismus, Wahrheit und Zeitlichkeit, Ein Beitrag zum Problem der zukünftigen kontingenten Wahrheiten', *Studia Leibnitiana* 6, (1974).
3 Łukasiewicz's views on this topic were formed, it seems, in the early 1920s. Their fullest expression is probably his essay 'O Determiniłmie' published posthumously in 1961. It has since been translated into English, in *Polish Logic 1920–1939*, ed. McCall (OUP. Oxford, 1967) and German, in *Studia Leibnitianà* 5 (1973).
4 By N. Rescher in his *Essays in Philosophical Analysis* (Pittsburg University Press, Pittsburg 1969), p. 274.
5 See J. Hintikka, *Time and Necessity*, p. 148.
6 When quoting Aristotle in English I shall here throughout use Anscombe's translation in 'Aristotle and the Sea Battle', *Mind* (1956), 1–15.
7 G. E. M. Anscombe, *op. cit*, p. 8: 'ἤδη, logical, not temporal: ἤδη works rather like the German "schon" (only here of course it would be "noch nicht")'. A translation with 'noch nicht' is all right I should say—but precisely because of its (vaguely) temporal connotation. See also J. Hintikka, *op. cit*., p. 174.
8 The otiose character of 'it is true that' is sometimes qualified by saying that the phrase is *semantically* otiose. One is then thinking of the fact that the phrase may have what is called a *pragmatic* function. By prefixing the phrase to a declarative sentence one may intimate that, in the context of saying or writing this, one is making a *statement*.
9 If the subject term of the sentence refers to a perishable individual, such

as Socrates, then *t* should be a time within the life-span of this individual.

10 See the beautiful passage in the *Nicomachean Ethics* 1139b7–11: 'no one deliberates about the past, but about what is future and capable of being otherwise, while what is past is not capable of not having taken place; hence Agathon is right in saying:

> For this alone is lacking even to God,
> To make undone things that have once been done.'

(I am quoting the translation by Sir David Ross.)

11 See, for example, *Analytica Priora* 34b7–9 where ἁπλῶς is contrasted with κατὰ χρόνον ὁρίσαντας.

12 If necessity and omnitemporal truth are identical then it follows logically that something is possible if, and only if, it is not omnitemporally false, i.e., is sometimes true. This 'definition' of possibility became later associated with the name of the Megarian Diodorus Cronus. Hintikka has in a number of publications with great acumen and vigour argued that this was also Aristotle's view of possibility. My own, amateurish, opinion is that Aristotle comes very close to this view, that it almost forces itself upon him, but that he was never trapped into a definite commitment to it. This, incidentally, seems to agree with Hintikka's more guarded attitude in his so far latest contribution to Aristotelian scholarship (written in collaboration with U. Remes and S. Knuuttila, 'Aristotle on modality and determinism', *Acta Philosophica Fennica* 29 (1977).

[16] COMING TRUE[1]

R. C. Jeffrey

> I mean, for example: it is necessary for there to be or not to be a sea-battle tomorrow; but it is not necessary for a sea-battle to take place tomorrow, nor for one not to take place—though it is necessary for one to take place or not to take place.[2]

In interpreting this sort of talk the sticking point is the familiar one concerning the truth value of $ApNp(=p\vee\sim p)$ when p is a future contingency. Here the temporal indexicality of 'There will be a sea-battle tomorrow' (said on Hallowe'en, 1984) is beside the point. The problem of future contingency also arises for 'There is (tenselessly) a sea-battle on 1 November 1984', which is no less a future contingency than is the indexical version, on the day before. The difficulties concern temporal indexicality in the metalanguage: Indexicality of 'p is (now) true', not of p itself. Here, then, I suppose that the object language is a fragment of English in which the atomic sentences are tenseless and non-indexical. A fuller account, treating tensed, indexical, and quantified sentences of the object language, requires complications of interest in their own right that leave untouched the problem treated here.

In his commentary, Ackrill[3] suggests 'ineluctably' as a substitute for 'necessarily' in Chapter 9 of *De Interpretatione*, for example in 19[a] 23, 'What is, necessarily is, when it is':

> In saying that it is not necessary that p is true Aristotle does not mean what we should mean if we said that a proposition is not a necessary truth but a contingent one. For he would say that it *is* necessary that p is true if the present state of affairs makes it certain that the p-event will occur, or again if the p-event has already occurred.

Then with $p=$ 'There is a sea-battle on 1 November 1984' I write $Ip=$ 'Ineluctably, there is a sea-battle on 1 November 1984', and I note that $p\vdash Ip$ (p implies Ip) but $Ip\nvdash p$ (that is, not conversely) for if the p-event has not yet taken place then p is not yet true, even if it is already ineluctable. This is in contrast to the familiar sense of necessity (read the box as 'necessarily') for which we have $\Box p\vdash p$ for all p but $p\nvdash\Box p$ for contingent *p*, for example for the *p* we have been considering. I first saw this point made by G. E. M. Anscombe, in 'Aristotle and the sea battle'[4] where she summarizes:

> Thus Aristotle's point (as we should put it) is that 'Either *p* or not *p*' is always necessary, and this necessity is what we are familiar with. But—and this is from our point of view the right way to put it, for this is a novelty to us—that when *p* describes a present or past situation, then either *p* is necessarily true, or $\sim p$ is necessarily true; and here 'necessarily true' has a sense which is unfamiliar to us. In this sense I say it is necessarily true that there was not—or necessarily false that there was—a big civil war raging in England from 1850 to 1870; necessarily true that there is a University in Oxford; and so on. But 'necessarily true' is not simply the same as 'true'; for while it may be true that there will be rain tomorrow, it is not necessarily true. As everyone would say: there may be or may not.

That struck me as gibberish when I first read it, some twenty years ago, but I cannot now recapture that sense of bafflement. The change came all at once, in 1975, when I began to think about the way in which we use conditionals in deliberation and in other situations where antecedent and consequent are not eternal truths or falsehoods ('Whales are mammals', '0 = 1') but are statements that have yet to assume their truth values, as for example in the following inference noted by P. T. Geach in *Reason and Argument*:[5]

> If Jim and Bill both turn their keys at 6:00, the missile will fire just after.
>
> ---
>
> If Jim turns his key at 6:00 the missile will fire just after, or if Bill turns his at 6:00 the missile will fire just after.

Now where *Cpq* is the truth functional conditional we do have *CKpqr*⊢*ACprCqr* (*K* for *Konjunction*), but I think that Geach's example shows that we want another conditional, too, for the English inference is invalid by the canons of ordinary practical reason, i.e., reasoning about how we might affect matters.

Here I shall suggest that *IC* works where *C* does not, as an interpretation of the conditional in such cases as Geach's missile example. The suggestion sounds rather like the one that Sextus Empiricus attributes to Diodorus Cronus:[6]

> Philo says that a sound conditional is one that does not begin with a truth and end with a falsehood . . . But Diodorus says it is one that neither could nor can begin with a truth and end with a falsehood.

Thus during or shortly after Aristotle's lifetime, Philo of Megara proposed the truth functional reading of the conditional, whereas Diodorus proposed a reading a bit like *IC*. But not quite: for one thing, Diodorus seems to have required the ineluctability to have held throughout the past ('neither could nor can'), as might be appropriate for a very strong sort of counterfactual conditional; but the *IC* reading is intended as forward-looking ('if *p is or comes* true') rather than counterfactual ('if *p were or had been* true').

For contrast with *I*, it will be useful to introduce another operator: *F* ('finally'). I take it that on Hallowe'en, *p* (as above) has neither truth value. To indicate that, I say that *p* has the value *u* on Hallowe'en. But if *p comes* true on 1 November, *Fp* has the value *t* ('true') on Hallowe'en, whether or not it was ineluctable then. Thus, the temporal indexicality of the present notion of truth is of no account for sentences of form *Fp*: each such sentence has one of the truth values, *t* or *f*, at all times, and has the *same* truth value at all times. But *Ip*, although it always has a truth value (*t*, *f*), may change its truth value from *f* to *t* (but never in the other direction). It will do so when the sea battle becomes inevitable, if it was not always so: If *Fp* is true, *Ip* will be true when *p* comes true, at the latest.

Underlying the present model theory is the conceit that the world grows by accretion of facts—to put the matter in the material mode of speech. But here I use 'world' as a term of art, and adopt the formal mode of speech. My worlds are more properly called 'notional worlds'. They are certain sets of things I call 'stages' (in the development of notional worlds), and stages are certain sets of sentences of the object language. Stages do duty (in the formal mode of speech) for all the facts so far. If an atomic sentence has already come true, it belongs to the present stage of the (notional world that corresponds to the) real world. If it has already come false, its denial belongs to the present stage of the . . . real world. In general, a truth functional conditional *Cpq* belongs iff (if and only if) *Np* does or *q* does or both, and the denial *NCpq* of such a conditional belongs iff *p* and *Nq* both do (so that falsity of the conditional is already settled). But neither of the operators *F* or *I* appear in sentences of stages: *the sentences in stages are to be baldly factual, and neither forward-looking* (*F*) *nor modal* (*I*).

Given a set of non-indexical atomic sentences, we define L as the set consisting of those together with all such sentences as can be obtained from them via the three unary operators F, I, N, and the binary operator C. (Other truth functional compounds are defined as usual, for example Apq as $CNpq$, Kpq as $NCpNq$, etc.)

L^0 will be the set of sentences in L that are free of F and I: the atomic sentences together with all and only the sentences obtainable from them via N and C. We define stages as Hintikka *model sets* of sentences of L^0: The *stages* are the sets S satisfying the following five conditions for all p, q in L^0.

(1) S is a (possibly empty) subset of L^0.
(2) At most one of p, Np is in S.
(3) NNp is in S iff p is.
(4) Cpq is in S iff Np is or q is.
(5) $NCpq$ is in S iff p is and Nq is.

We define *worlds* as sets w satisfying these three conditions:

(6) w is a set of stages (non-empty, by (8) below).
(7) If S and T are in w then $S \subseteq T$ or $T \subseteq S$.
(8) For each p in L^0 either p or Np belongs to some stage in w.

The inclusion relation in (7) identifies the temporal order of stages in the development of world w: S is earlier than T iff S is a subset of T but not vice versa (and, of course, $S = T$ iff $S \subseteq T$ and $T \subseteq S$). Condition (8) ensures that each world w is *complete* in the sense that the merger or union $\bigcup w$ of all stages of w contains the answer (p or Np) to every question of form 'Fp?' for p in L^0. Condition (2) ensures that we shall never get both answers to such a question.

We define *models* as non-empty sets of worlds. If W is the set of all worlds and $\varnothing$ is the empty set, the definition goes:

(9) M is a model iff $\varnothing \neq M \subseteq W$.

The interesting models will be *proper* subsets of W: $M \neq W$. The idea is that the worlds in M are to be those counted as *possible* (in a sense that the model theory leaves open). In part, the aim is to provide a framework within which one can represent various sorts of determinism and indeterminism by suitable choice of models.

Now we define inductively the *values* (t, f, u) assumed by sentences of L at stages S in the development of worlds w in models M. For p in L^0 this value, $V(p, S, w, M)$, is determined by the composition of stage S itself:

(10) For p in L^0, $V(p, S, w, M)$ is t, f, or u (*sc.*, undefined) according as p, Np, or neither is in S.

(Notice that V is being treated as a *partial* function with $\{t, f\}$ as its range. It is to be understood in (10) and similar contexts that $V(p, S, w, M)$ is undefined unless $S \in w \in M$.) For all p, q in L we then define the values of Np, Cpq, Fp, and Ip, as follows. In (11) the three-valued truth-tables are chosen so that when p and q are in L^0, the values assumed by Np and Cpq agree with those determined by (10) and (1)–(5).

(11) $V(Np, S, w, M)$ and $V(Cpq, S, w, M)$ are determined as functions of $V(p, S, w, M)$ and $V(q, S, w, M)$ by the tables below.

	N
t	f
f	t
u	u

C	t	f	u
t	t	f	u
f	t	t	t
u	t	u	u

Since $V(Fp, S, w, M)$ and $V(Ip, S, w, M)$ are never u, we give the conditions under which they are t with the understanding that where those conditions fail, the value is f.

(12) $V(Fp, S, w, M) = t$ iff $V(p, T, w, M) = t$ at some stage T of w and $V(p, U, w, M) = t$ at all subsequent stages $U \supseteq T$ of w.

(13) $V(Ip, S, w, M) = t$ iff $V(p, T, x, M) = f$ at no stage $T \supseteq S$ of any world x in M that has S as a stage.

For p in L^0 the *definiens* in (13) is equivalent to the requirement that Fp be true in every world in M of which S is a stage. And for such p (as Michael Frede has pointed out) the restriction '$\supseteq S$' in (13) does no work.

Being non-indexical, sentences of L^0 keep their truth-values once they assume them. Then for p in L^0, the last clause of (12) does no work: Fp is true in w iff p is true at some stage of w. But the operator I imports indexicality, so to speak: with p in L^0, Ip can come true, having been false, and therefore NIp can change its truth value from t to f. In such a case I want to say that NIp is *finally* false even though it was *sometimes* true. Thus, without its final clause ('and at all subsequent stages'), (12) would give F the sense of 'sometimes', not of 'finally'. The point is illustrated in Fig. 16.1 where lines are worlds, points are stages, and higher is later. Since NIp is true at stage S of either world (w or x) but false at stage T of w, we want to say that $FNIp$ is false in w, that is, false at every stage of w in M, even though NIp is true at stage S of w in M.

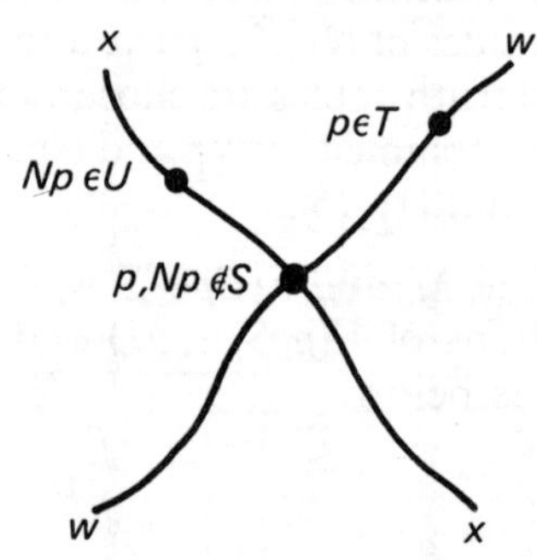

Fig. 16.1

In this framework, the strength of the operator I depends on what subset of W (viz., the set of all worlds) M is. At one extreme is the plenum. If $M = W$ and p is in L^0 then $V(Ip, S, w, M) = t$ iff $V(p, S, w, M) = t$, and 'ineluctably' has the force of 'it is already true that': there is no inevitability, that is, no forward-looking ineluctability for propositions in L^0. Then W is the totally *indeterministic* model.

At the other extreme, unit sets are totally deterministic models: if M contains just one world, I has precisely the force of F. The platitude 'What will be, will be' can then be strengthened to 'What will be is already ineluctable'. But determinism is more aptly understood as a property of worlds-in-models: world w is deterministic in model M iff disjoint from every other world in M (that is, iff w shares no stages with other worlds in M). From this point of view, the deterministic *models* are those in which no two worlds share any stage: models whose diagrams are sheaves of non-intersecting lines. If such a model contains two or more worlds, no one world *must* have been actual but each world is no less deterministic than it would have been had no others been possible.

An interesting class of models is that in which two worlds which share a stage must share all earlier stages. Where (as in the diagram above) worlds w and x share a stage S but do not share all earlier stages, the differences in the ways in which they develop toward stage S can only arise from differences in the order in which sentences true in S come true in the two worlds. In *tree models* (where diagrams that branch at all branch up but never down) such differences are ruled out. Definition:

(14) Tree models are models in which any two worlds that share a stage share all earlier stages.

If L^0 is rich enough, one might think it part of some concept of possibility that all models must be tree models.

It may be thought a defect of the present model theory that in it Cpp need not be true at every stage of every world in every model: Cpp need not be *valid*. (The reason is that where p is as yet undetermined, so is Cpp.) And it may be thought a defect that p may be undetermined where ineluctable. But it seems to me that validity of $ICpp$ answers to our intuitions about tautologies: in general, Ip is valid whenever p is a two-valued tautology. And it seems to me that validity of $CIpFp$ answers to our sense that p is true (*sc.*, finally true) where ineluctable. No sentence in L^0 is valid. Within L^0 the distinguishing mark of the two-valued tautologies is not validity (universal truth) but universal nonfalsehood, (for example we have $V(Cpp, S, w, M) \neq \mathrm{f}$ at every stage S of every world w in every model M).

For $\Delta \subseteq L$ we define validity of the inference from Δ to q as truth of q whenever all sentences in Δ are true:

(15) $\Delta \vdash q$ iff $V(q, S, w, M) = t$ whenever $V(p, S, w, M) = t$ for all p in Δ.

Writing '$\vdash p$' for '$\varnothing \vdash p$' as usual, '$\vdash p$' means that p is valid. And writing '$p \vdash q$' for '$\{p\} \vdash q$' as usual, we see that as usual, validity of the conditional Cpq implies validity of the inference from p to q, for we have

$\vdash Cpq$ iff $V(Cpq, S, w, M) = t$ everywhere,

$$V(p,S,w,M) = f \vee V(q,S,w,M) = t \text{ everywhere, i.e.}$$

$$V(p,S,w,M) \neq \text{f} \supset V(q,S,w,M) = t \text{ everywhere, while}$$

$p \vdash q$ iff $V(p,S,w,M) = t \supset V(q,S,w,M) = t$ everywhere.

(Here, 'everywhere' means: for all $S \in w \in M \subseteq W$.) But the converse does not hold: we can have $p \vdash q$ without $\vdash Cpq$, for example when p is atomic and $q = p$, there will be stages of worlds in models at which p has not yet assumed its truth-value, so that Cpq is u, and yet q is true wherever p is. With inference defined as in (15) we can have $p \vdash q$ but $Nq \nvdash Np$, for example for atomic r we have (and want) $r \vdash Ir$ but $NIr \nvdash Nr$: there will be stages of worlds of models at which r is not inevitable (so that NIr is true) and has not yet assumed its truth-value (so that Nr is not true).

Then (15) seems apt, although (inevitably) 'nonclassical'.

Now let us have a look at Geach's missile inference, with IC doing duty for *if*. The inference in question is invalid:

$$ICKpqr \nvdash AICprICqr$$

Indeed, in any model corresponding to the beliefs of one who takes the firing system to be failsafe (neither key alone can fire the missile) and who thinks it possible that the failsafe feature will be put to the test (so that there are states of worlds in which the system is operational, in which one key is turned without the other), there will be states of worlds in which neither component of the conclusion is true, but the premise is true (in other words, turning both keys would fire the missile). The simplest model illustrating the invalidity of this inference consists of just three worlds: $\{\varnothing, S\}, \{\varnothing, T\}, \{\varnothing, U\}$. L has just three atomic statements: p, q, r. S is the unique Hintikka model set containing p, q, and r; T is the one containing p, Nq, and Nr; and U is the one containing Np, q, and Nr. The tree diagram is shown in Fig. 16.2. At state $\varnothing$, $ICKpqr$ is true because $CKpqr$ belongs to all three of $\varnothing$'s possible future states. But at state $\varnothing$, $AICprICqr$ is false, that is, $ICpr$ and $ICqr$ are both false at $\varnothing$, for there is a possible successor (T) at which Cpr is false, and there is one (U) at which Cqr is false.

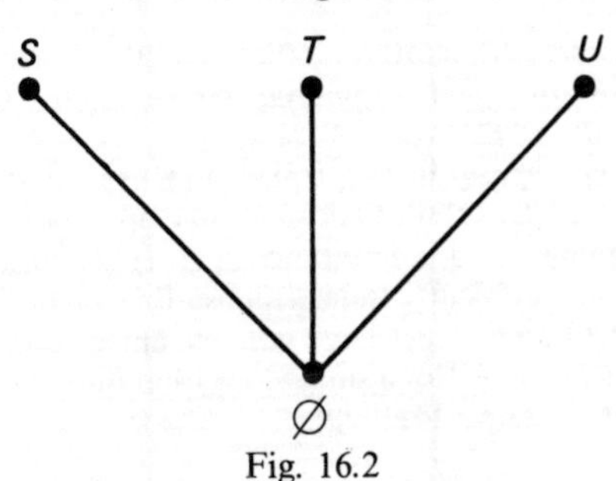

Fig. 16.2

It would seem that our ideas of objective probability might find a natural expression by converting the present models into probability models: having depicted the possibilities via tree structures, we assign non-negative numbers that sum to 1 as probabilities to the various paths leading from each node. That is how it would go in such simple cases as we have pictured here, anyway.[7]

Notes

1 Earlier versions of this paper had flaws other than those that remain, and I am grateful to those who called them to my attention, notably, F. C. Benenson. A version that is superceded by this one appeared in *Papers on Logic and Language* ed. D. Holdcroft, (University of Warwick, Department of Philosophy, 1977). Remaining here is a detail faulted by D. H. Mellor: condition (7), according to which the relation $\subseteq$ (is *earlier than or simultaneous with*) fully connects the stages of any one world. Mellor points out that this Aristotelian and Newtonian assumption is relativistically unacceptable, and H. Stein has suggested replacing (7) by the following weaker condition:

> (7′) If S and T are in w then for some U in $w, S \subseteq U$ and $T \subseteq U$. ('filtering to the right')

The present treatment will require structural changes to accommodate temporally indexical statements in L^0 and quantifiers in L^0. Even without such accommodations, it is unclear that (13) is the happiest definition of I. Two variants of (13), that come to the same thing as (13) for p in L^0, were noticed above. But those, and other plausible variants, differ from (13) for indexical p.

2 Aristotle, *De Interpretatione*, trans. *J. L. Ackrill* (*OUP, Oxford*, 1963), *Ch*. 9.

3 *Ibid.* p. 139.

4 G. E. M. Anscombe, 'Aristotle and the sea battle', *Mind*, 65 (1956), 1–15.

5 P. T. Geach, *Reason and Argument* (University of California Press, California, 1976), p. 92.

6 *Pyrrhoneiae Hypotyposes* ii, 110, trans. W. Kneale and M. Kneale, *The Development of Logic* (OUP, Oxford, 1962), pp. 128–129.

7 This paper was conceived in ignorance of R. H. Thomson's much earlier treatment of the problem, 'Indeterminist time and truth-value gaps', *Theoria* 36 (1970) and here I have not attempted to make contact with that or subsequent work. For a survey, see John Burgess, 'Logic and time', *Journal of Symbolic Logic* 44 (1979).

[17] MODALITIES IN NORMAL SYSTEMS CONTAINING THE *S5* AXIOM

Brian F. Chellas

The point of this paper is to prove the following theorem:

> Every normal system of modal logic containing the $S5$ axiom has finitely many distinct modalities.

A system of modal logic based on propositional logic (PL) is *normal* if and only if it contains the theorem $\Diamond A \leftrightarrow \neg \Box \neg A$ and is closed under the rule of inference:

$$\frac{(A_1 \wedge \ldots \wedge A_n) \rightarrow A}{(\Box A_1 \wedge \ldots \wedge \Box A_n) \rightarrow \Box A} (n \geq 0).$$

The following rules and theorems, present in every normal system, are of importance further on:

RM. $\dfrac{A \rightarrow B}{\Box A \rightarrow \Box B}$

RM$\Diamond$. $\dfrac{A \rightarrow B}{\Diamond A \rightarrow \Diamond B}$

T1. $(\Diamond A \rightarrow \Box B) \rightarrow (\Box A \rightarrow \Box B)$

T2. $(\Diamond A \rightarrow \Box B) \rightarrow (\Diamond A \rightarrow \Diamond B)$

T3. $\Box(A \leftrightarrow B) \rightarrow (\Box A \leftrightarrow \Box B)$

T4. $\Box(A \leftrightarrow B) \rightarrow (\Diamond A \leftrightarrow \Diamond B)$

By the $S5$ *axiom* I mean, indifferently, either

5. $\Diamond A \rightarrow \Box \Diamond A$

or

5$\Diamond$. $\Diamond \Box A \rightarrow \Box A$.

Because principles of duality hold in every normal modal logic, a system of this sort contains 5 just in case it contains 5$\Diamond$.

Normal systems containing the $S5$ axiom are also called

normal K5-systems – $K5$ being the smallest such system. The best-known normal $K5$-system is $S5$, which is a proper extension of $K5$. (For example, $S5$ contains $\Box A \rightarrow A$ and $A \rightarrow \Diamond A$, neither of which is a theorem of $K5$.)

Normal systems can also be characterized semantically, in terms of models. A *standard model* is a structure $M = \langle W, R, P \rangle$ in which W is a set (of 'possible worlds'), R is a binary relation in W, and P is a valuation determining which atomic sentences are true at which points in W.

Truth at a point α *in a model* M is defined in the customary ways for truth-functional compounds. For modal sentences we have:

> $\Box A$ is true at α in M if and only if A is true at every point β in M such that $\alpha R \beta$.
>
> $\Diamond A$ is true at α in M if and only if A is true at some point β in M such that $\alpha R \beta$.

A system of modal logic is said to be *determined* by a class of models just in case the theorems of the logic are exactly the sentences true at every point in every model in the class. Every class of standard models determines a normal system, and every normal system is determined by a class of standard models. Thus normal systems can be characterized as sets of sentences determined by classes of standard models.

Normal systems containing the $S5$ axiom are determined by classes of *euclidean* standard models—i.e. by classes of models satisfying the condition that, for every α, β, and γ, if $\alpha R \beta$ and $\alpha R \gamma$ then $\beta R \gamma$. In particular, $K5$ is determined by the class of all euclidean models.

Finally, a *modality* is a finite (possibly null) sequence of the operators $\neg$, $\Box$, and $\Diamond$, classified as *affirmative* or *negative* according as the number of occurrences of the negation sign is even or odd. Two modalities φ and ψ are *equivalent* in a logic if and only if the schema $\varphi A \leftrightarrow \psi A$ is a theorem; otherwise they are *distinct*. By well-known manipulations, licensed in any normal system, every modality can be put in the form φ or the form $\neg\varphi$, where φ is devoid of $\neg$.

This much is by way of a preliminary. We are ready now to prove the theorem.

For the proof it is enough to show that every normal $K5$-system contains the following 'reduction laws':

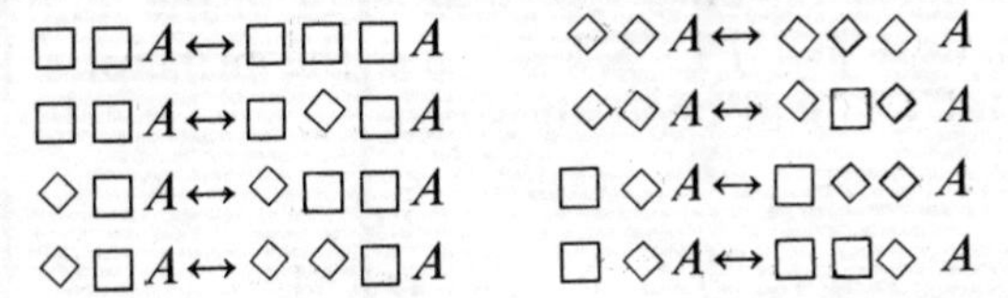

$$\Box\Box A \leftrightarrow \Box\Box\Box A \qquad \Diamond\Diamond A \leftrightarrow \Diamond\Diamond\Diamond A$$
$$\Box\Box A \leftrightarrow \Box\Diamond\Box A \qquad \Diamond\Diamond A \leftrightarrow \Diamond\Box\Diamond A$$
$$\Diamond\Box A \leftrightarrow \Diamond\Box\Box A \qquad \Box\Diamond A \leftrightarrow \Box\Diamond\Diamond A$$
$$\Diamond\Box A \leftrightarrow \Diamond\Diamond\Box A \qquad \Box\Diamond A \leftrightarrow \Box\Box\Diamond A$$

These laws might be summed up by the motto 'delete the middle modality'. In virtue of these theorems every modality reduces to one containing at most two occurrences of the operators $\Box$ and $\Diamond$. Thus every normal $K5$-system has at most fourteen distinct modalities—to wit, the seven diagramed in Fig. 17.1, where • is the null modality, and their negations. (The arrows in Fig. 17.1 indicate the implications among the affirmative modalities; for their negations, reverse the arrows.)

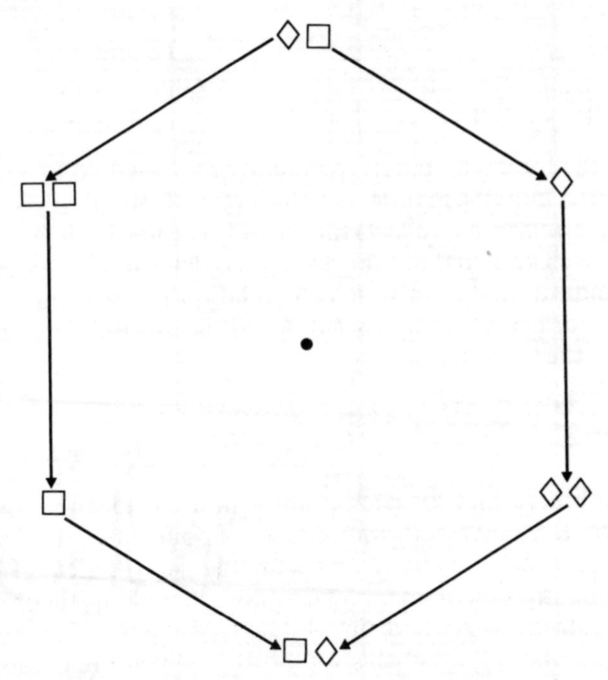

Fig. 17.1. Modalities in normal *K5-systems*.

The reduction laws on the right follow from their mates to the left by duality. For the four on the left the requisite conditionals appear on lines 2, 3, and 9–14 of the following proof (which uses just *PL*, 5, 5◇, and the principles RM, RM◇, T1, and T2 mentioned above).

1.	$\Diamond\Box A \rightarrow \Box A$	5◇
2.	$\Box\Diamond\Box A \rightarrow \Box\Box A$	1, RM
3.	$\Diamond\Diamond\Box A \rightarrow \Diamond\Box A$	1, RM◇
4.	$\Diamond\Box A \rightarrow \Box\Diamond\Box A$	5
5.	$\Diamond\Box\Box A \rightarrow \Box\Box A$	5◇
6.	$\Diamond\Box A \rightarrow \Box\Box A$	2, 4, *PL*
7.	$\Box\Diamond\Box A \rightarrow \Box\Box\Box A$	6, RM
8.	$\Diamond\Box A \rightarrow \Box\Box\Box A$	4, 7, *PL*
9.	$\Box\Box A \rightarrow \Box\Diamond\Box A$	4, T1 and *PL*
10.	$\Diamond\Box A \rightarrow \Diamond\Diamond\Box A$	4, T2 and *PL*
11.	$\Box\Box\Box A \rightarrow \Box\Box A$	5, T1 and *PL*
12.	$\Diamond\Box\Box A \rightarrow \Diamond\Box A$	5, T2 and *PL*
13.	$\Box\Box A \rightarrow \Box\Box\Box A$	8, T1 and *PL*
14.	$\Diamond\Box A \rightarrow \Diamond\Box\Box A$	8, T2 and *PL*

To the foregoing purely syntactic proof I would like to add another, more semantic in character. I think it is more illuminating, and it reflects the way I first came to the theorem.

It is well known that the system *S*5 is determined by the class of standard models in which the relation is universal—i.e. in which, for every α and β, $\alpha R\beta$. And it is also well known that *S*5 has these reduction laws:

$$\Box A \leftrightarrow \Box\Box A \qquad \Diamond A \leftrightarrow \Diamond\Diamond A$$
$$\Box A \leftrightarrow \Diamond\Box A \qquad \Diamond A \leftrightarrow \Box\Diamond A$$

Now observe that for each point α in a euclidean model the relation R is universal within the set of points β such that $\alpha R\beta$. That is, $\gamma R\delta$ for every γ and δ in the set $\{\beta: \alpha R\beta\}$. By the determination theorem for *S*5 it follows that all the theorems of *S*5 are true everywhere in $\{\beta: \alpha R\beta\}$. In particular, the *S*5 reduction laws hold in this set. So by the truth definition the necessitations of these schemas:

$$\Box(\Box A \leftrightarrow \Box\Box A) \qquad \Box(\Diamond A \leftrightarrow \Diamond\Diamond A)$$
$$\Box(\Box A \leftrightarrow \Diamond\Box A) \qquad \Box(\Diamond A \leftrightarrow \Box\Diamond A)$$

are true at every point in every euclidean model. Hence by determination they are theorems of every normal *K*5-system. But then so, too, are the eight reduction laws, by *PL* and the theorems T3 and T4 mentioned above.

It is easy to describe euclidean countermodels to show that there are no further reduction laws in *K*5. Thus fourteen is precisely the number of distinct modalities in the smallest normal system containing the *S*5 axiom.

The theorem and corollaries for several extensions of *K*5 are proved in my forthcoming book.[1] I cannot say for certain that the result is new, but I believe it is. Ideas and information pertinent to the proofs of the theorem can also be found in material written by E. J. Lemmon and Dana Scott[2] and by Krister Segerberg.[3] It should be noted, however, that terminology and notation in these texts differ somewhat from mine.

Notes

1 B. F. Chellas, *Modal Logic: An Introduction* (CUP, Cambridge, forthcoming).

2 E. J. Lemmon and D. Scott, *An Introduction to Modal Logic* (*The 'Lemmon Notes'*), in *American Philosophical Quarterly Monograph Series*, No. 11, ed. K. Segerberg (Basil Blackwell, Oxford, 1977).

3 K. Segerberg, *An Essay in Classical Modal Logic* (University of Uppsala, Uppsala, 1971).